T0402232

The Economics of Evaluating Water Projects

Per-Olov Johansson • Bengt Kriström

The Economics
of Evaluating Water Projects

Hydroelectricity Versus Other Uses

 Springer

Prof. Per-Olov Johansson
Stockholm School of Economics
Department of Economics
Stockholm
Sweden

Prof. Bengt Kriström
Umeå University and SLU
Centre for Environmental
and Resource Economics
Umeå
Sweden

ISBN 978-3-642-27669-9   e-ISBN 978-3-642-27670-5
DOI 10.1007/978-3-642-27670-5
Springer Heidelberg Dordrecht London New York

Library of Congress Control Number: 2012934149

Printed on acid-free paper

Springer is part of Springer Science+Business Media (www.springer.com)

Preface

The main purpose of this book is to present research on a kind of water use conflicts that probably will become more and more common and important as time goes by. In times of increasing demands for electricity as well as environmental services the question arises how to best managing moving water. Should more be diverted to or from electricity generation? That is the kind of timely question which this book addresses. We develop a simple general equilibrium model of a small open economy which is used to derive a cost-benefit rule that can be used to assess projects that divert water from electricity generation to recreational and other uses (or vice versa). The cost-benefit rule is used to evaluate a proposed change at a Swedish hydropower plant. According to the proposal, water is diverted from electricity generation to the natural river channel (dryway) creating environmental and recreational benefits.

We present a framework for assessing the costs and benefits of small perturbations of current hydropower regulatory regimes. The framework integrates several key issues, including, but not limited to: a contract between the hydro power plant and another party (local residents) generating the general equilibrium cost-benefit rule; the contract is a corner stone of our referendum-style contingent valuation study; the tax system in the status quo; (partial) foreign ownership of the plant; trade in electricity; trade in renewable energy certificates; trade in carbon emission permits; externalities of replacement power (generated in other countries); value of loss of regulating (balancing power) and other system services; transmission of electricity modeled as provided by natural monopolies; downstream hydrological externalities and environmental benefits (aesthetic, recreational and otherwise). We also present a contingent valuation study where local residents participate in a hypothetical referendum about the proposal. The web-based questionnaire has some novel features with respect to the format of the valuation questions, introducing an open-ended willingness to pay question plus an interval question within a hypothetical local referendum. We also put some numbers on the different items of the cost-benefit rule in order to illustrate possible magnitudes of different benefits and costs. A quite extensive sensitivity analysis is provided introducing new tools that we believe will be useful in future cost-benefit analysis.

The research presented in this book was carried out as a part of the R&D programs PlusMinus – Economic Assessment for the Environment – sponsored by the Swedish Environmental Protection Agency and Hydropower – Environmental Impacts, Mitigation Measures and Costs in Regulated Waters – financed by Elforsk, the Swedish Energy Agency, the National Board of Fisheries and the Swedish Environmental Protection Agency. Many people from these organizations have contributed in discussions of different aspects of hydroelectricity generation. We are most grateful for these contributions.

Several persons have helped us with the contingent valuation study: Scott Cole, Kjell Leonardsson, Bo Ranneby, and Peter Rivinoja. Yuri Belyaev and Ying-Fu Xie kindly helped us with some statistical computations. Magnus Ekström has been extremely generous in providing comments and suggestions on various mathematical statistical issues. We are also most grateful to Mr. Kent Pettersson, Fortum, for kindly providing us with estimates of the revenue loss of the hydropower plant at Dönje caused by the scenarios considered in this book. Andrew Zaeske made the manuscript readable by scrutinizing the language. Participants in our conference on CBA and Hydropower in Stockholm on August 26, 2009, provided us with many insights and stimulating comments and suggestions. However, all estimates presented in this book are based on own calculations. Therefore, any remaining errors are our own responsibility.

Finally our thanks to Annual Reviews – A Nonprofit Scientific Publisher – for permitting the use of parts of the article "The New Economics of Evaluating Water Projects" written by the authors of this book and published in Volume 3, 2011, of Annual Review of Resource Economics. The extracted parts of the article constitute Chap. 2 of this book. Similarly, we thank the Berkeley Electronic Press for permitting the use of the article "Comment on Burgess and Zerbe: On Bank Market Power and the Social Discount Rate" published in Volume 2, 2011, of Journal of Benefit-Cost Analysis.

Stockholm and Umeå   *Per-Olov Johansson*
                                                                        Bengt Kriström

Contents

List of Tables

List of Figures

Chapter 1
Introduction

Regulating a river alter flows and generally has strong negative impacts on most aquatic organisms and on those in the riparian zone (e.g. [159]), as well as detrimental effects on recreational possibilities and aesthetic values associated with rivers. On the other hand, hydropower offers energy free from emissions. In addition, hydropower is extremely cost-effective in countries with suitable natural conditions, like Norway and Sweden. Taken together, these facts suggests a conflict between competing uses of river resources in general and between policy objectives in particular. Consider the European Water Framework Directive, which formalizes the demand for improved ecological status of water bodies within the union in terms of quantified minimum levels. Simultaneously, the union has unleashed its "triple 20 by 2020" policy, which includes reducing carbon emissions by 20%. Furthermore, economic growth is a perennial policy objective and energy demand typically goes hand in hand with growth. These facts maps into (increasingly sharper) conflicts about the proper husbandry of our water resources, conflicts that somehow must be resolved.

Cost-benefit analysis[1] (CBA) offers a formal approach to delineating the costs and benefits of different policies and may provide useful information for decision-making. Quite arbitrarily, we refer the reader interested in reading more about the theoretical principles of project evaluation to [45, 100] and [140]. These manuals are quite formal and demand some knowledge of general equilibrium theory. However, there are also many cookbook style manuals providing detailed advice on how to proceed in a real-world application, see e.g. [40, 50, 86, 149], and [49]. An introduction to the underlying welfare theory is found in [99] while more technical presentations are provided by [20] and [140].

Because the origins of empirical CBA can be traced back to dam constructions in the U.S. in the 1930s, there is a very substantial body of relevant literature to tap. While the "older" literature did not include detailed benefit assessments, we now

---

[1]In Europe the approach is typically denoted cost-benefit analysis while in the U.S. it is often denoted benefit-cost analysis. We will throughout follow the European tradition in this respect.

P.-O. Johansson and B. Kriström, *The Economics of Evaluating Water Projects*, DOI 10.1007/978-3-642-27670-5_1, © Springer-Verlag Berlin Heidelberg 2012

have a substantial number of studies that attempts to quantify the various non-priced benefits associated with a diversion of water from hydropower production. Samples from the U.S. include studies on the recreational benefits of salmon fishing in rivers [94, 127, 146], and nonuse value associated with restored rivers and recovered salmon populations [10, 168].[2]

Our CBA includes the benefits of a changed instream flow. In an early survey of this field, [126] summarizes the findings. The reported values varies widely (about USD 0.5 (see [141]) to USD 74 per acre foot). Important factors affecting the values are e.g. the amount and the timing of water. Reference [81] surveyed lakefront property owners, recreational users and potential users of a reservoir to analyze how property values and recreational values would be affected by changing the level of water by 2 in. in Lake Koshkonong in Wisconsin. Reference [46] use contingent valuation (see Chap. 5 for details about this approach) to find the value of changed instream flows in the Montana's Big Hole and Bitterroot Rivers. Marginal recreation values per acre-foot at low flow levels are between USD 10 and USD 25.

There also exists a substantial literature on the cost side. For example, hydroelectric projects in the U.S. require license renewals, which often include compulsory fish passage improvements. In recent years, a number of major hydroelectric projects in the Northwest have considered the costs of such upgrades (see e.g. [53]) while government agencies have assessed the economic impacts of protecting the endangered salmon and bull trout populations [144, 174]. In addition, economic studies have examined the cost-effectiveness [92, 93] and total expenditures [145] of measures to improve up and downstream migration for fish. Reference [139] examines cost effectiveness of various measures to reduce temperature in salmon watersheds. Other studies have examined the costs and impacts of artificial fish propagation [91].

A number of in-depth studies have examined the costs (e.g. foregone revenue) and benefits (use and non-use values) associated with maintaining minimum flows in regulated rivers; the reader is referred to Sect. 5.1 for definitions of use and non-use values. Those studies are very similar in spirit to ours. For example [41] estimated USD 325 million in costs (1994 present value dollars) associated with purchasing replacement peak power sources when hydroelectric facilities draw down reservoir levels slowly to assist fish migration. Reference [67] examine alternative "water banking" schemes that allow parties to purchase water for the purpose of assisting endangered species. Similarly [95], estimate the costs for policy makers to purchase water to ensure minimum flows for fish. Finally [14], estimate the non-use values associated with increased instream flows for the purpose of protecting endangered fish species in the southwestern U.S.

---

[2]The classic study is [32]. There is a rather extensive similar literature in the Nordic countries. Indeed, the hydropower expansion in Norway was subject to intensive analysis using various methods to describe the consequences. Reference [148] has a summary of the literature on the value of recreational fishing, including a survey of Nordic studies. In a recent CBA of hydropower-salmon conflicts in Sweden, [72] presents a detailed analysis of changing the water regime to the benefit of wild salmon.

Our study differentiates itself from the mentioned studies in several ways. We develop a small open economy general equilibrium cost-benefit rule which is used to evaluate small changes in power production by companies that are (totally or partially) owned by foreigners. The core of the rule can be viewed as a contract specifying the (minimum) compensation paid to a company in lieu of a specified reduction in its level of production. In addition, our rule specifies how to handle tax distortions and indicates how to put monetary values on some important externalities. This framework integrates several key issues, including, but not limited to:

- A contract between the hydropower plant and another party (local residents) generating the general equilibrium cost-benefit rule, which is a corner stone of our referendum-style Contingent Valuation study
- The tax system in the status quo
- (Partial) foreign ownership of the plant
- Trade in electricity, renewable energy certificates[3] and carbon emission permits
- Externalities of replacement power (generated in other countries)
- Value of loss of regulating (balancing power) and other system services
- Transmission of electricity modeled as provided by natural monopolies
- Downstream hydrological externalities
- Environmental benefits (aesthetic and otherwise).

A rather unique aspect of our empirical analysis is that we have assessed the environmental impacts from the perturbation by an actual experiment at the plant. Thus, the perturbations have been implemented in a "test-run" and the ecological consequences thus monitored "live" on-site. We proceed by providing a brief description of the scenarios with more details to follow in a later chapter.

1.1 The Dönje Power Plant Scenario

Dönje power station is a 67 megawatt[4] (MW) hydroelectric facility on the Ljusnan river close to the small city of Bollnäs in central Sweden. The city of Bollnäs has around 13,000 inhabitants and the municipality of Bollnäs has around 26,000 inhabitants. The power plant is part of a 28 power plant system spread out over the Ljusnan and its small tributary Voxnan. Ljusnan, in mid/northern Sweden, is regulated along most of its length for hydropower production. The river originates from the mountains at the Norwegian border and has a catchment area of 19,800 km^2. It is 440 km long and has an annual mean flow of 226 m^{3}s^{-1} at the

---

[3] Also known as renewable electricity certificates.

[4] According to [125]. Its normal annual production is 340 gigawatt hours (GWh) according to www. Kuhlins.com.

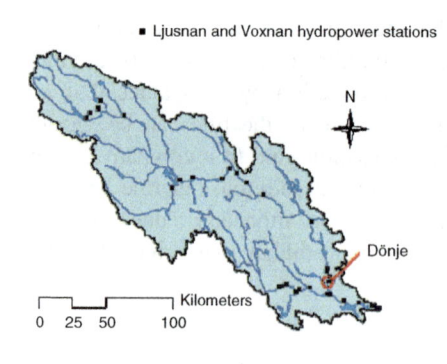

Fig. 1.1 The Dönje power plant

mouth (N61°12, E17°08) in the Bothnian Sea.[5] The plant and the catchment area is depicted in Fig. 1.1.

The river section Bollnäsströmmarna (N61°22, E16°24), c. 6.5 km in length, is located between power-station five and six in order from the river mouth, about 53 km upstream of the sea. Generally most of the water reaching this section is utilized by the power-station Dönje that has a maximum capacity of $250\,\mathrm{m}^3\mathrm{s}^{-1}$. Here the hydraulic head is created by 34 m deep intake tubes to turbines after which the water is lead via a tunnel to the outlet in the downstream Lake Varpen.

Since water is diverted from the natural river stretch this result in reduced flows in Bollnäsströmmarna. According to the hydropower licence conditions the minimum flow in the original stretch is to be $10\,\mathrm{m}^3\mathrm{s}^{-1}$ from May 15 to October 21, after which the spill is gradually decreased to $0.5\,\mathrm{m}^3\mathrm{s}^{-1}$ for the winter period (October 31–May 14). In summer most of the water is spilled into the 1.3 km long eastern river branch, Klumpströmmen, that gains $0.25\,\mathrm{m}^3\mathrm{s}^{-1}$ in winter, while the western branch is more or less dry for most of the year. Both scenarios that we will evaluate in this book target Klumpströmmen, see Fig. 1.2 for a map.

Before regulation about one third of the total flow amount in Bollnäsströmmarna passed at the Klumpströmmen section, while the remaining 2/3 passed the western river branch. The natural flow regimes showed the typical patterns for rivers in northern Sweden; the highest flow events occurred in spring (peaking c. $740\,\mathrm{m}^3\mathrm{s}^{-1}$), with decreasing and stabilizing flows in the summer (normally c. $15\text{–}70\,\mathrm{m}^3\mathrm{s}^{-1}$), occasionally with an increase in autumn, and relatively low constant flows during the ice-period in winter (averages around $10\,\mathrm{m}^3\mathrm{s}^{-1}$). Returning to the current situation, in practice the dryway is more or less bottom frozen during winter, and the downstream stocks of fishes that are valuable for fishing are small. The scenic view when the river channel is dry is far from overwhelming. Recreational activities

---

[5]Several of the numbers used in this section to describe the scenario and the area are lifted from a project companion report, focussing the natural science measurement at the plant, see [154].

Fig. 1.2 Map of
Klumpströmmen at the power
plant. Note: "Torrfåra" = dry
river bed, "Generellt
Fiskeförbud" = Fishing not
allowed

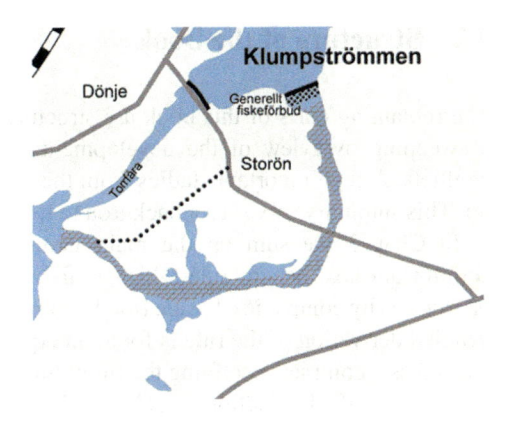

other than fishing such as canoing and ice skating are also adversely affected by the
regulation.

In the Contingent Valuation study two scenarios were introduced, based on focus
group studies, and in-depth discussions between various stakeholders, ecologists
and economists. According to the first scenario the winter flow would increase
from $0.25\,\mathrm{m^3s^{-1}}$ in Klumpströmmen to $3\,\mathrm{m^3s^{-1}}$ while the summer flow remained
unchanged at its current level of $10\,\mathrm{m^3s^{-1}}$.[6] In the second scenario we also increased
summer flow, from 10 to $20\,\mathrm{m^3s^{-1}}$. These scenarios are denoted SCENARIO 1
(*winter*) and SCENARIO 2 (*winter-summer*), respectively. It should be noted that
both scenarios entail relatively small perturbations of the current regulatory regime.
As a first step to measure the benefits we used a small web-questionnaire, which
is described briefly in Chap. 5. Additional details about the scenario and how we
assessed the costs are contained in Chap. 4.

Finally, we note the following curiosity. The legal process, which began in the
later part of the 1940s, was complex and extensive. The summary document of all
legal processes up until today is 2,652 pages long. In a letter dated 15 March 1950
from the "Naturskyddsföreningen", the largest environmental Non-Governmental
Organization (NGO) in Sweden, its secretary notes that the minimum winter and
summer flows should be increased to 1 and $3\,\mathrm{m^3s^{-1}}$, respectively.[7] The company
had asked for a minimum release of 1 and $0.25\,\mathrm{m^3s^{-1}}$. The secretary, Nils Dahlbeck,
considered the preferred levels necessary to avoid still standing water. He went on to
produce a legendary TV show on the environment and is well-known in the minds
of many Swedes.

---

[6]From an ecological point of view the optimal flow mimics the natural flow. A "natural flow"
scenario was considered but not implemented in our study due to its complexity.

[7]Svenska Naturskyddsföreningens yttrande ang. Dönje kraftverk vid Bollnäs, letter to Advokat-
fiskalsämbetet vid Kungl Maj:ts och Rikets Kammarkollegium, dated 15 march 1950, on file at
Bollnas municipality archives, Box EIV a:4.

1.2 Structure of the Book

The remaining parts of this book are structured as follows. In Chap. 2 we present a sweeping overview of the development of cost-benefit analysis of water use conflicts. A few important studies from the United States and Europe are summed up. This summary serves as a background to our own work.

In Chap. 3 we sum up the main items of a small open economy general equilibrium cost-benefit rule which is used to evaluate small changes in power production by companies that are (totally or partially) owned by foreigners; the more detailed derivation of the rule is found in Appendix A. The core of the rule can be viewed as a contract specifying the (minimum) compensation paid to a company in lieu of a specified reduction in its level of production. In addition, our rule specifies how to handle tax distortions and indicates how to put monetary values on some important externalities.

In the empirical part of the book we use the cost-benefit rule to assess small reductions in the level of electricity generation at the hydroelectricity plant at Dönje in Sweden. Water is diverted from electricity generation to the natural river channel (dryway) according to the two scenarios described previously, which creates environmental and recreational benefits.

In Chap. 4 we then discuss how to assess the loss of profits as the level of production at Dönje is reduced according to the two considered scenarios. There are many tricky issues to address, including how to estimate the physical loss of electricity as water is diverted to other uses, and how to forecast the electricity spot price and how to select a societal discount rate.

The non-priced items of the cost-benefit analysis are discussed in Chap. 5. In particular, we present a Contingent Valuation study where local residents participate in a hypothetical referendum. The web-based questionnaire has some novel features with respect to the format of the valuation questions since we introduce an open-ended willingness to pay (WTP) question plus an interval question.

In Chap. 6 we put some numbers on the different items of the cost-benefit rule in order to illustrate possible magnitudes. We label the results point estimates since there is a single number for each item. We also present quite a few ways of undertaking sensitivity analysis, including stochastic approaches that we believe are novel. However, it must be stressed that our CBA is too tentative to be used in decision making. It is part of an ongoing project that will ultimately provide more accurate estimates of the different items of the CBA.

Finally, Chap. 7 contains a few concluding remarks. Appendix A has details about the model, Appendix B contains the equations used to compute present values for the cost-benefit analyses, Appendix C lists spot electricity prices, Appendix D describes pertinent features of the web-questionnaire, Appendix E provides the data used to calculate WTP, and Appendix F has the computer code for most of the econometrics. We have used R [151], a freely available program, so that our results can conveniently be replicated. The estimates behind the actual cost-benefit analysis are fully replicable.

Chapter 2
Origins of CBA of Water Use Conflicts*

Although the conceptual underpinnings of CBA may be traced to the nineteenth-century French civil engineer and economist Jules Dupuit, extensive application of the method had to wait until the twentieth century.[1] The sharpening of CBA into a potentially useful decision-making tool also involved engineers, i.e. the U.S. Army Corps of Engineers (ACE). Propelled by a rising demand for electricity and substantial damage from several serious floods, the U.S. Congress passed two significant flood control acts (1936, 1944) (referred to here as the 1936 Act and the 1944 Act, respectively).[2] The 1936 Act called for "works of improvement" on more than 50 major rivers throughout the United States and made flood control a federal government activity. The necessary physical constructions had been the ACE's expertise, and ACE became heavily involved in many construction projects, as it was given responsibility for analysis of rivers for flood control (whereas the U.S. Department of Agriculture was given responsibility for water flow on upstream watersheds.[3])

Importantly, the Act introduced an approach to prioritizing between projects:

> The federal government should improve or participate in the improvement of navigable waters or their tributaries, including watersheds ...for flood-control if the benefits to whomsoever they may accrue are in excess of estimated costs (as quoted in [68])

---

* This chapter draws extensively on [105].

[1] The Swedish water law of 1918 stipulates that no one is allowed to build a hydropower plant unless the revenues after deduction of the construction cost is at least twice the damage inflicted. It goes on to stipulate that construction is not permitted if a fishery is seriously affected. Nor is construction permitted if there are very serious ecological damages. In an abstract sense this law suggests that a hydropower plant must pass a cost-benefit test. Furthermore, the law stipulates compensation for damages, implying the Pareto criterion rather than Kaldor-Hicks (since compensations were actually paid); again, this is in the abstract, but compensations potentially had to be negotiated for each stakeholder in the vicinity of the proposed plant. The process that led to the construction of the Dönje plant provides an example.

[2] According to [112], the federal Reclamation Act of 1902 required economic analysis of projects.

[3] For further details on the problems created by this separation of tasks, see [153].

P.-O. Johansson and B. Kriström, *The Economics of Evaluating Water Projects*, DOI 10.1007/978-3-642-27670-5_2, © Springer-Verlag Berlin Heidelberg 2012

These ideas were later developed in several handbooks and manuals. The 1944 Act gave the Corps responsibility for multi-purpose dams, e.g. hydropower construc-tions. [4] CBA subsequently conquered new worlds and new applications in the 1950s and onwards, as the tool was applied to various types of public projects in Europe and later on in the third world countries.

The classical contributions that can be said to have sprung out of the 1936 Act include [47, 82, 83, 120], and [131]. Of these, [131], a key output from the Harvard Water Program, stands out as the most comprehensive in terms of perspective, as it presses a system perspective and multiobjective planning. Thus, sophisticated computer modeling was used to generate insight into large sets of alternatives, using – somewhat controversially – a multiple objective framework. This work was further developed by [90] who demonstrated the use of simulation modeling within a detailed application of the methods to the Lehigh River system (and, in less detail, the Delaware river system). The simulation model is exceptionally detailed, including a hydrological projections module, descriptors of water supply and demand and a range of development measures. In spirit, this systems approach to water project is reminiscent of the EU Water Frame Directive, in that the Directive emphasizes a river basin perspective (a perspective that is also important in [120]).

Rather than providing a detailed review of the empirical literature that exists, we discuss a few high-quality contributions, mainly from the US. In addition, we will illustrate some recent ideas in the application of CBA to water projects, as they flow from our framework. While the problems are encountered world-wide, we will concentrate on U.S. and Europe. Thus, we begin with the U.S., then turn to review a study in Sect. 2.2 on applying CBA to the Water Framework Directive and one study applying a dynamic approach to cost-benefit analysis.

2.1 The United States

We target studies that more or less directly are related to energy from moving water.[5] We begin with a capsule summary of the U.S. literature and then turn to the classical U.S. studies. This brief review goes some way in appreciating the difficulties encountered in the pioneering studies. Indeed, we have come quite a long way, especially regarding the measurement of benefits. It is possible that the attitude, in some cases, towards benefit measurement was affected by the paucity of such studies. We will give an example below, but let us now turn to the facts.

---

[4]For a historical review of the development of CBA in the U.S. the reader is referred to [189], for Australia to [43], and for some UK studies to [75]. A fine book-length treatment of water planning in the U.S. that covers the developments in great detail is the volume by [157], see, especially, Table 3.1 in the chapter by [21] for a comprehensive overview of the developments.

[5]For a recent summary of valuation studies on water quality, see e.g. [15].

Reference [128] has collected 1,239 estimates on the value of 30 outdoor recreation activities, a subset of which is relevant for water projects. He concludes that hunting on public lands in the Pacific Northwest has an average value of $35 per day, fishing averages $42 per day, wildlife viewing $35 per day, and hiking $24 per day in the Pacific Northwest. He reports on about 20 studies of river recreation such as rafting, kayaking, and canoeing in the U.S., which yield about 80 separate valuation estimates. The average value of these recreation activities is $101 per day, with a 95% confidence interval of $82–$120. The highest average values per day are for the Southeastern U.S. at $127 per day. Values for unique sites are even higher, at $300 per day. Reference [22] has performed a meta analysis on fishing recreation studies, suggesting that the value of an angler day is $11.63 higher for fishing days taking place in rivers than in standing freshwater (e.g. reservoirs, ponds). For further details on additional studies (e.g. property value studies for dam removals and passive use values for preserving existing free flowing rivers), we refer to [129]. Finally [14], estimate the non-use values associated with increased instream flows for the purpose of protecting endangered fish species in the Southwestern U.S. Reference [164] is an extensive early study of the benefits of water quality.

A number of in-depth studies have examined the costs (e.g. foregone revenue) and benefits (use and non-use values) associated with maintaining minimum flows in regulated rivers. For example [41] estimated USD 325 million in costs (1994 present value dollars) associated with purchasing replacement peak power sources when hydroelectric facilities draw down reservoir levels slowly to assist fish migration. Reference [67] examine alternative "water banking" schemes that allow parties to purchase water for the purpose of assisting endangered species. Similarly [95], estimate the costs for policy makers to purchase water to ensure minimum flows for fish.

As indicated above, applied CBA arose from the efforts in the U.S. to shed light on the benefits and costs of water projects. In one of the classic studies, [120] looked at private- and public-sector activities in river basin development from an welfare economics point of view. They explained the role of welfare economics and the tools that can be used in practice to measure benefits and costs. In the same year, another classic study [47], was published. The book brought together the existing approaches towards benefit measurement through the lens of formal welfare economics. As summarized by [111] (p. 46),

> A common feature of all of them [studies on "conventional" outputs of water projects] was that benefits could be satisfactorily evaluated by ingenious applications of information generated by markets. Eckstein's *Water Resources Development* was an exposition and critique of these methods and an interpretation of them in terms of formal welfare economics.

However, the early studies largely ignored recreational values since methods to measure them were not yet available. For example [47] claims that "recreation must be judged on other criteria, for the use of benefit-cost analysis for them not only is invalid, but casts general doubt and suspicion on procedures which can effectively serve a high purpose where they are appropriate" (p. 41).

Reference [119] forwards an early critique of the view that environmental benefits lack an economical value. He was concerned with more or less unique natural environments that would be permanently destroyed if developed, so developing such environments or resources would cause irreversible effects or damages. One of Krutilla's arguments is that the willingness to pay for developing a resource is limited to the area under a demand curve for the resource's services (if we take a static perspective here, for simplicity). On the other hand, the minimum which would be necessary to compensate losers is not limited to such areas. Reference [119] concludes that it is impossible to determine whether the market solution is efficient or inefficient. At first glance it seems a bit surprising that he does not connect his discussion to property rights and the Coase theorem; see [34]. However, the explanation is that Krutilla was dealing with situations where no property rights could be established. Another of Krutilla's arguments draws on the concept of option value when demand is uncertain; see [180] for details. Even if a person does not currently consume a particular good he or she might value the option to consume it in the future. Thus retaining the pure possibility of future consumption has a value.

Reference [119] did also consider natural environments as endowments for certain groups. His argument opened up for the possibility that resource stocks are included as arguments in individuals' utility functions (as in the framework above). Then stocks as well as harvests are modeled in utility maximization problems, although from an individual's point of view a stock's path over time might be exogenous. The "conventional" (instantaneous) utility function $u[c(t)]$, where $c(t)$ is consumption at time t is replaced by a function $U[c(t), s(t)]$, where $s(t)$ denotes the stock of a natural resource at time t with $s(0) = \bar{s}$ denoting the initial and given endowment of the resource. In the simplest case the path of the stock evolves according to the equation $\dot{s}(t) = f[s(t)] - h(t)$, $f[s(t)]$ is the natural growth of the resource (if any), and $h(t)$ refers to the harvest of the resource. This way of modeling the problem, which is only sketched here, opened up for analysis of the welfare significance of national product in a dynamic economy as is indicated by the title of the classical article by [181]. Early contributions with a focus on how to augment the conventional net national product measure so as to become a better welfare indicator in the presence of natural resources are [37, 80], and [132].[6]

The following quotation neatly summarizes Krutilla's are views of the state of the art of cost-benefit analysis in the 1960s:

> Accordingly, our problem is akin to the dynamic programming problem which requires a present action (which may violate conventional benefit-cost criteria) to be compatible with the attainment of future states of affairs. But we know little about the value that the instrumental variables may take. We have virtually no knowledge about the possible magnitude of option demand. And we still have much to learn about the determinants of the growth in demand for outdoor recreation and the qualitative significance of the asymmetry in the implications of technological advances for producing industrial goods on the one hand

---

[6]For reviews and extensions, see [4] and [84].

and natural environments on the other. Obviously, a great deal of research in these areas is necessary before we can hope to apply formal decision criteria comparable to current benefit-cost criteria (p. 785).

Fewer than 10 years later [121] published another major contribution, "The Economics of Natural Environments".[7] This book addresses several important issues. In contrast to [119, 121] discuss the Coase theorem. They also model an intertemporal decision problem involving an irreversible investment constraint. We do believe this is one of the earliest applications of optimal control theory in environmental and resource economics. They also discuss the concept of quasi-option value, a concept attributed to [5, 56], and [85]. Finally [121] discuss how to empirically measure the benefits of preserving a natural environment,[8] acknowledging that they do not fully understand how to estimate such benefits. Nevertheless they suggest an approach which can be useful whenever we lack detailed and robust benefits estimates.[9]

The Coase theorem states that in a competitive economy with complete information and zero transaction costs the allocation of a resource will be efficient independently of the initial allocation of property rights. For example, the optimal level of pollution of a river will be the same regardless of whether polluters, say farmers, or fishermen are assigned property rights. These rights determine the direction in which compensation is made if rights are violated. However the invariance part of the theorem holds only if preference relations are quasi-linear so that payments cause no income effects. This is also noted by [121]. A simple way of stating this is by noting that the willingness-to-pay (the compensating variation) for access to a resource differs from the willingness-to-accept compensation (the equivalent variation) in exchange for access unless the considered individual's preference relation is quasi-linear. In addition the informational requirement is strong as is the assumption of zero transaction costs, in particular when a large number of individuals are involved.

To handle investment decisions under an irreversibility constraint they use optimal control theory. They maximize the discounted net benefits:

$$\int_0^\infty e^{-rt}[b^d(s(t),t) + b^p(s(t),t) - C(I(t))]\mathrm{d}t \tag{2.1}$$

subject to $\mathrm{d}s/\mathrm{d}t = I$, $I \geq 0$, and $s(0) = \bar{s}(0)$. In Eq. 2.1 $b^d(.)$ refers to the *net* benefits at time t of developing a resource (and the interest rate at which items are discounted is denoted r). These are a function of the scale of development or the size of the project $s(t)$ and might also change over time as indicated by the time argument t. There is an opportunity cost of development covered by $b^p(.)$.

---

[7] A revised edition of the book was published in 1985 where, in particular, the option value issue was further developed.

[8] At the time, some economists argued that CBA did not apply if there were irreversible effects. See the discussion (and defense of CBA in such cases, referring to [121]) by [163].

[9] For application of their framework, see e.g. [74, 158] and [185].

For example, if development means building a dam recreation is possibly damaged. The scale of the project might be the number of dams constructed. Thus $b^d(.)$ is increasing in the scale of the project while $b^p(.)$ is decreasing in $S(t)$. Finally the flow capital cost of the project is denoted $C(I(t))$. The constraints are the dynamic constraint on development, the restriction on reversibility, and the initial condition indicating that the planner starts with an exogenously given project scale $\bar{S}(0)$at time zero. This so-called *Krutilla-Fisher model* has attracted much interest in the literature, and we will return to it in our own cost-benefit analysis .

Reference [121] focus on likely time patterns of $b^d(.)$ and $b^p(.)$. They find it likely that the benefits of development of a natural resource are decreasing over time. One reason being technological change over time. The benefits of preservation, on the other hand, are likely to increase over time. One reason being that technological change can hardly produce close substitutes to environmental resources while increasing incomes tend to increase the willingness-to-pay for environmental resources (if they are normal, as is typically assumed). Reference [162] develops these ideas in detail.

Reference [121] also discuss the concept of option value (although the concept dates back at least to [119], or even Stanley Jevons, as argued in [117]). There are several definitions available, but the two most common in environmental and resource economics are the real option value associated with Dixit and Pindyck, see, for example, [42], and the quasi-option value associated with Arrow-Fisher-Henry-Hanemann. The relationship between these two concepts has been analyzed (in a simple two-period model) by [137]. The Dixit-Pindyck real option value of postponing an investment, here denoted OV^{DP}, can be decomposed as follows:

$$OV^{DP} = OV^{AFHH} + PPV, \qquad (2.2)$$

where OV^{AFHH} is the quasi-option value (≥ 0) reflecting the value of information (the difference between closed-loop and open-loop solutions), and PPV is the pure postponement value, i.e. the value of having the option of postponing the investment. This last value, which is also present in a deterministic setting, might be positive or negative ($\lesseqgtr 0$). For example, if net flow benefits of a project are negative today but will become positive at some future time, it is profitable to delay the project until that point in time. Unfortunately, and as acknowledged by [121], estimating option values is difficult. Still the concept might be useful in circumstances where the known benefits of development are of roughly the same size as the benefits of preservation and development causes irreversible damage to preservation interests.

Let us add that another concept of option value was developed from Krutilla's early paper. This concept was ultimately recognized as a measure of risk aversion in a framework that characterized choice with state dependent preferences and with discrete characterization of uncertainty. Reference [65] is the paper that clarifies this literature.

Reference [121] undertake several empirical cost-benefit analysis, devoting two chapters to an impressive analysis of a hydro power development in the Hells

Canyon[10] where they estimate a slightly simplified version of Eq. 2.1. A lot of space is devoted to discussions how b^d and b^p might change over time. However, to today's reader it is obvious that they lack the instruments for estimating the willingness to pay for having the canyon preserved. In those days (the Hells Canyon study was undertaken in 1969) there was at least a travel cost study [31] and a study [39] using a survey technique that today is known as the Contingent Valuation method. Reference [121] refer to both but did not use these instruments in the Hells Canyon study. Travel cost methods can be used to estimate use values, or the willingness to pay of visitors or users of a resource. Passive use values relate to the value of the continued existence of a natural environments and their living organisms as well as bequest motives, such as a desire to also allow future generations to "consume" these resources. Passive use values cannot be measured in markets but require other methods such as surveys where respondents are asked about their willingness-to-pay for the preservation of a particular resource, i.e. a natural environment.

To the best of our knowledge, the first study of water use conflicts incorporating use values as well as passive use values is [173], which investigated alternative strategies to no action at the The Glen Canyon Dam, the second largest dam on the Colorado River. The dam's main purposes include generating electrical power, storing water for the arid southwestern United States, and providing water recreation opportunities. Construction of the dam began in 1956 and it was able to begin blocking the flow of the river in 1963. Due to the stability of the dam, there has not been the periodic flooding that would wash away and renew sand banks along the portion of the Colorado River that transits the Grand Canyon. Because of the stability of the sand banks, several non-native species of plants became established, adversely affecting the native wildlife (http://en.wikipedia.org/wiki/Glen_Canyon_Dam). The silt that once flowed and settled along the Colorado River streambed are now contained to Lake Powell reservoir. The reduction of silt in the water has been of great importance socio-culturally for the people and the animals that have had to learn to live with the consequences. Another cost of the Glen Canyon Dam is evaporation that takes place in Lake Powell (http://www2.kenyon.edu/projects/Dams/glen.html).

This environmental impact statement was completed in 1995; see [173]. It uses conditions as of 1990 as the baseline and evaluates eight alternative strategies to taking no action. The study aims at estimating use values as well as non-use values of the various alternatives. For both types of values the Contingent Valuation method is used. Reference [16] undertook a study of river-based recreation in the area. The study found that the value of angling and white-water boating was related to flow and that there were significant differences between the effects of flow on commercial

---

[10]The lower Snake River, forming the boundary between the northeastern border of Oregon and the west central portion of the Idaho border, passes through about 200 miles of a geologic formation known as Hells Canyon. The Snake River in this reach is one of the most scenic streams to be found anywhere according to [121].

white-water boaters and private white water boaters. However, the study was unable to find a correlation between the value of day use rafting and flow.[11]

Reference [173] attempted to cover non-use values using the Contingent Valuation method. A focus group in New Mexico and two subsequent focus groups in Arizona explored whether individuals held any value beyond a use value for the hydropower resource. No non-use value for the hydropower resource per se was evident. However, participants in these focus groups clearly empathized with particular populations, such as small farmers and rural residents, whose lifestyles might be affected by the price impacts associated with the loss of peaking capability at Glen Canyon Dam. Therefore, descriptions of residential price impacts and impacts on farmers were developed and included in the pilot test survey instruments.

A pilot study was mailed to a national sample and a sample of households living in the affected area. The results are reported in [183]. Pilot-test sample sizes were too small to allow for statistically reliable estimation of non-use value for each of the alternatives. Non-users were most concerned about impacts to vegetation and associated wildlife, native fish, Native Americans, and archeological sites. Therefore alternatives that benefit these resources are likely to have higher non-use value. Pilot-test results indicated that estimates of non-use value obtained in the full-scale study may range from tens to hundreds of millions of dollars annually. The results of the full-scale study were not available to the authors of [173]. However, a study by [184] contains a massive account of all steps towards the full scale survey of 8,000 households in the United States.

In any case we do believe that this highly impressive study (covering 320 pages excluding bibliography and appendices) is the first to estimate use values as well as non-use values of a major water use conflict.[12]

2.2 Europe

We next discuss a study on how CBA has been used to shed light on how the Water Framework Directive might affect the usage of water resources in Europe and turn then to a Swedish study that is used to illustrate certain analytical points in our framework. Note that none of our European studies apply the Krutilla-Fisher approach to discounting, i.e. assuming the benefits can be discounted differently from costs. It can be argued that the environmental resources in these particular

---

[11]Reference [173] also estimates what looks like a *Keynesian* spending multiplier. That is, every dollar spent in the considered region will generate further incomes that in part are spent, generating further incomes, and so on.

[12]However, since the results of the full-scale survey reported in [184] were unavailable, non-use values were not included in the cost-benefit analysis of the alternatives to the baseline scenario (no action).

cases are not unique and hence the same type of irreversibility argument does not apply. Again, we abstain from a detailed literature review.[13]

2.2.1 The Water Framework Directive and CBA

The Water Framework Directive entered into European law on 22 December 2000 and covers EU water bodies, including rivers, lakes, reservoirs, wetlands, groundwaters, estuaries and coastal waters. The Directive's objective is to maintain environmental quality and to achieve "good status" in ecological and chemical terms by 2015. Implementation is based on river basin management planning, a mechanism for integrated catchment management. Plans are implemented over a 6-year cycle, which includes characterization, identification of pressures and impacts, and the setting of environmental objectives and programmes of measures to allow them to be met. Good ecological status is the default ecological target for all waters, but a kind of exemption is allowed via the so-called Heavily Modified Water Bodies (HMWB) (Article 4). A number of criteria must be fulfilled in order for a water body to qualify as a HMWB, including the key idea in Article 4(3)(b)) of being "disproportionately costly". Thus, if it is considered "too costly" to reach the goals, the water body can be given a HMWB status and hence an exemption. This exemption rule has led to a significant demand for CBA, as well as controversy, as the HMWB is considered a "loophole" in some circles and "a logical necessity" in yet others.

Reference [73] provide an illustration of how CBA has been used under the WFD. The study focused two cases, including the river Tummel in Central Highlands, Scotland. This river basin was selected to shed light on damages caused by hydro-electric development. All water bodies in the Tummel system were identified as candidate Heavily Modified Water Bodies (HMWB). Reference [73] asked if improving the ecological status of the Tummel system implies costs that are "disproportionately" greater than benefits. Because removing all hydropower related structures in the Tummel basin was considered an infeasible option, [73] looked at alternative scenarios that allowed continued hydropower, while promoting ecological quality towards good ecological status. Compensatory flows released to the rivers immediately downstream of hydro control structures and an upgrading of fish passes to enhance their effectiveness were principal measures suggested. These measures were considered to be practicable for the Rivers Garry and Errochty which form part of the east of the Tummel catchment. The essential part of the cost was found to be the value of lost electricity production. Estimated benefits

---

[13]The interested reader is referred to www.evri.ca for an inventory, and [142] and [143] for detailed surveys of the European literature on valuation. Reference [148] has a summary of the literature on the value of recreational fishing, including a survey of Nordic studies. Similar to the U.S. studies mentioned above, he also reports significant variation of valuations across type of fish, conditions at the site, income and so on.

included valuing the return of migratory salmonids to the Rivers Garry and Errochty, because this would increase rental values (and hence revenues) on the River Tummel downstream of the study area. The authors acknowledge that there are several categories of benefits that have not been included. Yet, their conclusion is that the proposed options for reaching a Good Ecological Status are "disproportionately costly" and therefore that these water bodies should be designated HMWB. A key factor in arriving at this conclusion is also the additional fact that alternative ways of generating peak-demand electricity are more detrimental to the environment.

2.2.2 A Salmon-Hydropower Conflict: Dynamic Model

In a recent study [72] analyzed a salmon-hydropower conflict. Society is assumed to value both its consumption of a numéraire good, produced via hydropower, and its consumption of salmon. In addition, the society attributes a value to the stock of salmon. The good is produced in a hydropower plant that creates the trade-off we are interested in here: more electricity generation means less salmon and vice versa. We sketch this trade-off in the following manner (ignoring initial and terminal conditions):

$$\text{Max} \int_0^\infty u\left(c(t), c_L(t), s_L(t)\right) e^{-rt} dt \tag{2.3}$$

subject to

$$c(t) = f(\dot{s}_w(t)) \tag{2.4}$$

$$\dot{s}_w(t) = k(t) - A(t), \tag{2.5}$$

where $c(t)$ denotes consumption of the numéraire good at time t, $c_L(t)$ denotes consumption of salmon at time t, $s_L(t)$ denotes the stock of salmon at time t, and r is a discount rate, for simplicity assumed to be constant across time. The amount of electricity is a function of the flow of water $\dot{s}_w(t)$ through the power station, which is determined by natural conditions (precipitation, etc.), denoted $k(t)$, and the amount of water diverted from electricity production, denoted $A(t)$, to the salmon (i.e. diverted into the old riverbed) to help salmon pass the station.

The change of the stock of salmon is determined as follows:

$$\dot{s}_L(t) = g(s_L(t), A(t)) - c_L(t), \tag{2.6}$$

where $g(.)$ is a generic growth function. Thus the natural growth of the stock of salmon is a function of $A(t)$ and the size of the stock of salmon at time t and Eq. 2.6 shows that the change of the salmon stock at time t is equal to the natural growth of the stock at time t less the catch of salmon at time t. For simplicity, it is assumed that the catch of salmon at time t is equal to consumption of salmon at time t.

A cost-benefit rule is arrived at by considering $A(t)$ as a parameter and proceeding in the way detailed in, for example, [4]. The resulting dynamic cost-benefit rule was used in structuring a Contingent Valuation study. Data comes from a project regarding a potential salmon passage-hydropower conflict in the northern Swedish Ume River and its largest salmon producing tributary, the Vindel River, see [72] for details about this study. Daily water flow data is combined with daily data on the number of salmon (1974–2000) that pass the hydropower plant Stornorrfors. Detailed ecological studies were used to build a Contingent Valuation scenario and to study the opportunity costs of releasing more water versus the potential benefit of salmon upstream migrants. Two ecological models were developed, one predicted upstream migration success of salmon in the bypass channel at added bypass flows, another predicted the number of salmon passing the fish ladder after improved upstream migration. The ecologists used results from 1995 to 2005 of tracking radio-tagged upstream migration salmon in the Ume River, and 30 years of information from the fish ladder. The scenario entails reducing production of electricity, which would increase the number of wild salmon in the river, as more water would be allocated to salmon passage areas. The mail survey was carried out in the autumn of 2004, using a random sample of 3,200 Swedes above the age of 18. The response rate was 66%.

On average, approximately 3,000 wild salmon per year that reached the Vindel River's spawning grounds between 1995 and 2004. This figure was used as a baseline in the scenario. The forecast on future number of salmon was built on the ecological models, and given the uncertainty about how the salmon stock responds to a perturbation of the water regime, a number of different forecasts were used in different versions of the survey. The results suggested strongly that valuations are independent of the number of salmon in the interval 4,000–9,000 per year, implying an existence value. It appears that respondents would like to pay for increasing the stock, i.e. securing the salmon, a kind of existence value.

The key idea was to base the valuation question on the general equilibrium cost-benefit rule hinted at above. Therefore, a proper valuation-question should, according to the cost-benefit rule, give information about both the direct effects of the salmon stock due to a change of water flow diverted to salmon and the impact on growth of salmon due to an increase of water flow. It was assumed that the method used to increase the amount of salmon does not influence people's valuations, as long as it is considered a natural method (and the increase of salmon is the same regardless of the method). The study scrutinized preferences across different methods, in this case improving the fish ladder or increasing the water flow in the bypass channel.

In her analysis of the benefits and costs, [72] concludes that passive use (non-use) values are the major contributors to the benefit (SEK 96–517 m[14]) of increasing the wild salmon stock in the Vindel River. The sensitivity analysis suggests that the opportunity costs in terms of lost electricity are typically higher than the estimated benefits.

---

[14]In our own cost-benefit analysis a Swedish Krona (SEK) is assumed to correspond to 0.1 Euro (EUR).

Chapter 3
A Simple Cost-Benefit Rule

In this chapter we discuss how to design a cost-benefit rule to be used to assess the reregulations, i.e. changes in water use, suggested by the two considered scenarios. We present a simple general equilibrium cost-benefit rule for a tax-distorted economy. The small or marginal project under consideration diverts water from electricity production to more environmentally friendly uses. The items in the associated cost-benefit rule are discussed in this chapter but the more detailed derivations are contained in Appendix A. It should be added that our evaluation is *ex ante*, i.e. we consider reregulations that have not yet occurred; for a recent *ex post* analysis of dam relicensing in Michigan the reader is referred to [116].

3.1 The Basic Rule

The special feature of the project under consideration is that it involves a private, reasonably profit-maximizing multinational firm in part owned by foreigners. At first glance, the scenario seems inexpensive to Swedes, because a large fraction of the owners are not part of the Swedish society. A significant fraction of the loss of profit is borne by individuals outside the conventional definition of a society, unless Swedes have some altruistic reasons to include the well-being of these foreigners in their utility functions. However, if we respect property rights, there is no obvious way to force the firm to deviate from its profit maximizing use of its water use rights. Therefore, the way to proceed is to provide the firm with an incentive to use less water for electricity generation. In effect, this is equivalent to buying back some of the company's water use rights or to in some other way "bribe" the company to change its level of production.[1]

---

[1] From a legal point of view the Swedish water use rights concept is quite complex. In this book we will speak of buying/selling such rights. Thereby we simply mean a contract between two parties stating the present value sum of money paid in exchange for a specified change in water use.

P.-O. Johansson and B. Kriström, *The Economics of Evaluating Water Projects*, DOI 10.1007/978-3-642-27670-5_3, © Springer-Verlag Berlin Heidelberg 2012

In Appendix A we work through a simple general equilibrium model of a small open economy. Cost-benefit rules are generated by marginally changing a parameter and tracing the changes from one general equilibrium to another. The parameter used here is interpreted as a contract according to which the firm receives a sum of money in return for a reduction in its electricity generation at the Dönje plant. The resulting societal cost-benefit rule for a *small* change can be stated as follows,[2]

$$dW^M = d\pi_c^F + WTP^B + WTP^{RB} - WTA^E + dT^2, \tag{3.1}$$

where dW^M denotes the change in societal welfare converted to monetary units, $d\pi_c^F < 0$ denotes the before-tax loss of profits for the owner Fortum calculated at constant/initial prices, WTP^B denotes the aggregate or total national willingness-to-pay for a smoother downstream flow of water, WTP^{RB} denotes the aggregate national willingness-to-pay for obtaining an improved downstream river basin (exclusive of the willingness-to-pay for a smoother flow of water), WTA^E denotes the aggregate national willingness-to-accept compensation in lieu of increased emissions of climate and other gases e.g. sulphur or nitrogen, and dT^2 refers to changes in taxes as explained below. In what follows we will briefly consider each of the terms in Eq. 3.1.

An illustration of the profit loss argument is found in Fig. 3.1 where the (highly simplified) "supply ladder" shifts to the left due to the considered project; it is

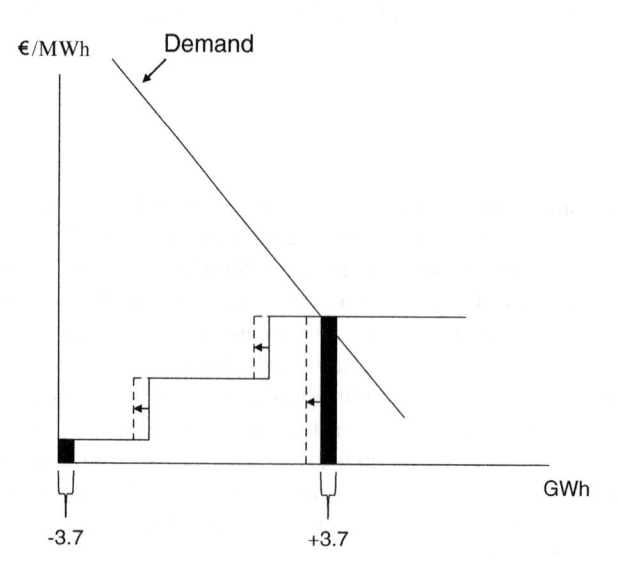

Fig. 3.1 A simplified spot market for electricity

---

assumed that the annual loss of electricity is equal to 3.7 gigawatt hours (GWh).
Cheap hydropower is replaced by electricity generated by a marginal supplier. The
principal cost for the project is the difference in costs between the two power
sources; the reader might note that the difference between the two dark staples
is equal to the annual loss of profits for the Dönje plant if its annual production
decreases by 3.7 GWh.

In an atemporal setting the pre-tax loss of profits is simply,

$$d\pi_c^F = [(p^s + p^b)dx^F - w(1 + t^w)d\ell^F], \tag{3.2}$$

where p^s is the spot price of electricity on the Nord Pool market,[3] p^b refers to the
value of regulating and system services provided by flexible electricity generators
like hydropower, $dx^F < 0$ denotes the loss of production at the Dönje plant, w
is the wage rate, t^w is a social security fee, and $wd\ell^F \le 0$ is the change in labor
cost associated with the reduction in output. For simplicity in Eq. 3.2 all inputs but
labor are ignored, although this is not the case in the empirical CBA. There are
stochastic short-run (within the hour) variations in supply (a nuclear power plant,
for example, might suddenly shut down) and demand (due to a sudden change in
temperature, for example) and also forecast errors by producers. Such variations are
largely handled by hydropower in the Nordic countries; the reader is referred to [59]
and [60] for detailed analyses of the properties of different regulating services. This
fact motivates that the considered plant is attributed a value over and above the spot
price, i.e. a total of $p^s + p^b$ per kilowatt hour (kWh). It should be underscored,
though, that only a small fraction of output is devoted to these regulating services.
However, the other side of the coin is that sudden changes in the water flow may
cause damages to the river basin and sometimes creates a moon-like landscape. If
the considered change at Dönje causes a less volatile or more smooth water flow,
there is a benefit. The WTP for such a change is captured by the term WTP^B in
Eq. 3.1. We make a distinction between WTP^B and WTP^{RB} since the Contingent
Valuation study attempts to cover only the latter term.

In the empirical study, we estimate intertemporal versions of Eq. 3.2. Such
variations takes the form,

$$d\pi_{NPV}^F = \int_0^\Gamma [(p^s(t) + p^b(t))dx^F - w(t)(1 + t^w)d\ell^F]e^{-rt}dt, \tag{3.3}$$

where Γ is the time horizon, and r is the social discount rate. The choice of r,
including the issue whether it should be constant or hyperbolic, is addressed in
a later chapter. Similarly, we will discuss possible trajectories for prices over the
considered time horizon. Moreover, the existing plant has a limited economic life.
Sooner or later it must be replaced by a new one. This raises the question if the size

---

[3]For details about this intra-Nordic market, see www.nordpool.com

of the plant (the dam, the turbines, and so on) can be marginally reduced if some water is diverted from passing through the turbines. Our basic assumption is that the investment cost is more or less unaffected by the kind of small changes in water use considered here. Therefore a variation of Eq. 3.3 with Γ set equal to 150 years is used in our base case CBA; this basically approximates an infinity assumption. However, in the sensitivity analysis we will return to the investment issue. It should be noted that we ignore the issue how to handle uncertainty in a CBA in this book except in our sensitivity analysis. In [107] we model the opportunities of an individual electricity producer as a stochastic optimal switching problem assuming that the price of electricity at time t and the price of the carbon permit at t are described by mean-reverting stochastic diffusions.

3.2   Further Details

As was described in the previous chapter, the main purpose of the project under consideration is to improve the recreational and other values of the basin downstream the Dönje power plant. The willingness to pay (WTP) for these values is captured by the term WTP^{RB} in Eq. 3.1. Since Fortum has acquired the water use rights it is legitimate to use a WTP concept rather than a willingness to accept compensation (WTA) concept.[4] In other words, the local residents must pay in order to obtain an improved environmental quality. As in Eq. 3.3 we must consider an intertemporal WTP measure. This will be further discussed later on.

In the empirical study it is assumed that $WTP_i^{RB} \geq 0$, where a subscript i refers to an individual living in the municipality of Bollnäs.[5] Since both scenarios analyzed in this study are so relatively small we set $WTP_j^{RB} = 0$ for all individuals j living outside the considered municipality. Thus in Eq. 3.1 WTP^{RB} is simply equal to the sum of WTP_i^{RB} for all i, i.e. all residents of the municipality of Bollnäs.

The willingness-to-pay for a smoother water flow, i.e. WTP^B, is discussed and estimated separately. It turns out to be so small that we set it equal to zero for both scenarios under consideration.

In our two scenarios a reduction in the electricity generation at Dönje is assumed to be covered by increased production by (foreign) fossil-fired power plants. This will be further clarified when we describe the operation of the Nord Pool spot market. These fossil-fired plants cause emissions of climate gases. However, if there are permit markets net emissions will remain unchanged, as is illustrated in Fig. 3.2. A fossil-fired plant that increases its electricity generation must acquire permits which means that through a price adjustment some other producer will be induced to reduce its emissions, i.e. to sell some of it permits. However, not all climate gases

---

[4]For discussion of these concepts the reader is referred to, for example, [98] or [23].

[5]In order to simplify notation, people living in nearby communities are included in the municipality of Bollnäs.

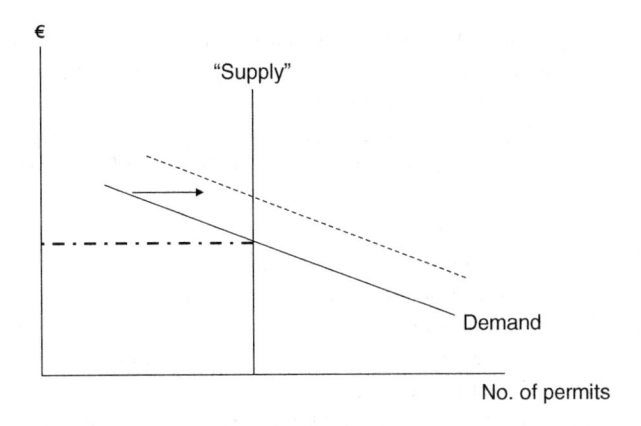

Fig. 3.2 A simple illustration of a permit market

are covered by the European trading scheme. Moreover, emissions of other gases, such as sulphur and nitrogen, might increase as Sweden imports more fossil-based electricity.[6]

A conventional cost-benefit analysis deals with monetary welfare consequences at the national level. The key question is if this implies that consequences of a project occurring outside the boarders of the country should be ignored. The answer is provided by the fact that a conventional cost-benefit analysis, just like conventional welfare theory, relies on the concept of consumer sovereignty, that individual preferences should be respected. Therefore, if Swedes are "nationalistic" egoists in the sense that they care only about damage within the boarders of the country the cost-benefit analysis should ignore any damage caused abroad by replacement power. On the other hand, if Swedes care about the impact of their actions irrespective of where the damage occurs, a cost-benefit analysis should respect this fact. This is also seen from the general equilibrium cost-benefit expression in Eq. A.17 in Appendix A. In the base case we will assume that the representative Swede is an altruist in the sense that he or she cares about any damage Swedish actions cause abroad.[7] The term WTA^E is supposed to cover the minimum aggregate (national) compensation needed for accepting the total of all

---

[6]The power plant acquiring permits must crowd out some other producer demanding permits. From a general equilibrium perspective it is more or less impossible to estimate the net impact on emissions. The outcome depends, for example, on whether a steel producer or a oil refinery is crowded out. Moreover plants might be reallocated from the Union to locations in other parts of the world.

[7]The reader is referred to [101, 108], and [114] for detailed discussion of different altruism-concepts but a brief overview is found in Sect. 5.1 below.

extra emissions, regardless of whether they cause damage domestically or abroad.[8] It is important to realize that we here consider a WTA concept rather than a WTP one. The reason is that an increase in emissions causes a loss of welfare. Thus the interpretation of Eq. 3.1 is that a necessary condition for the project to be recommended is that WTP^{RB} at least amounts to $WTA^E - d\pi_c^F - WTP^B$ (ignoring here the tax term dT^2).

An alternative to the national level approach used here would be to use a European Union or even a global perspective where the costs to and benefits for the Union or the globe are estimated but such an approach seems to be at odds with the conventional definition of a social cost-benefit analysis which loosely restricts it to dealing with a nation or country. However, there are exceptions, for example, the famous "Stern Review" of the global costs of climate change, see [167], and [57] who discuss evaluations at the European Union level. It is also important to stress that this book throughout considers small or marginal projects. For recent general equilibrium evaluations of large or non-marginal projects, the reader is referred to [29, 30], and [124].

Finally, there is a tax term in Eq. 3.1, which, in fact, is very similar to the general equilibrium tax rule stated in Eq. 15 in [64]. As is further explained in Appendix A within the simple framework considered in this book the term in question has two elements. The first component refers to a change in demand for electricity, and for simplicity we ignore the theoretical possibility that demand for other goods might change. If demand changes as a result of the considered project, government's revenues from the unit tax on electricity and value added tax will change as follows,

$$(t^{el} + t^{vat} p^{BV}) dx^d, \tag{3.4}$$

where t^{el} is the unit tax on electricity, t^{vat} is the value added tax (VAT) as a proportion of the before-VAT consumer price p^{BV} of electricity, and dx^d is the change in demand (ignoring here other sources of electricity demand than household demand). If demand decreases by the same magnitude as production at Dönje we should value the cost not at producer price but at a "partial" consumer price p^{pa},

$$p^{pa} = p^s + t^{el} + (p^s + t^{el} + \alpha p^c + p^{tr}) t^{vat}, \tag{3.5}$$

where p^s is the spot price, α is the proportion of demand that according to Swedish law must be provided by renewable sources, p^c is the price of energy certificates[9] (needed for α percent of demand), p^{tr} is the variable part of the price for transmission including regulating services, and t^{el} is the unit tax on electricity consumption. We speak of a "partial" consumer price since the prices paid for

---

[8]Even a hydropower plant causes emissions in a life cycle perspective. However, since we only consider a marginal change in water use we assume that there is a net increase in emissions if we shift to fossil-fired plants.

[9]As far as we understand the Dönje plant is not qualified for receiving energy certificates.

energy certificates plus transmission are not present in Eq. 3.5, except through the value added tax term. In the case of energy certificates we assume that the (equilibrium) price is necessary in order to induce the marginal provision of "green" energy, i.e. that price just covers the marginal cost. With respect to transmission we assume natural monopolies with decreasing average costs.

In Appendix A we present a simple model of a natural monopoly using two-part tariffs with a fixed and a variable part and a break-even constraint. According to this model it might be socially optimal to have variable tariffs that exceed marginal cost, since the firm is otherwise unable to survive. Hence, according to our model of the grid we should ignore any difference between p^{tr} and the marginal cost in transmission when estimating (3.5); Eq. 3.5 reflects the tax wedge on electricity.

The result that we should value at a partial consumer price is obtained by using Eqs. 3.2 and 3.4 with $dx^d = dx^F$ in Eq. 3.1. Note that in this case $WTA^E = 0$ since there is no replacement of the electricity generation lost, assuming that the change in production at Dönje does not affect emissions. We will consider this case, i.e. $dx^d = dx^F$, in the sensitivity analysis but in the base case the assumption that $dx^d = 0$ is maintained. Hydropower plants in Sweden have to pay a property value tax which is constructed in such a way that it often is viewed as a production tax. This tax is ignored here since it is paid out of sales revenues and hence represents a pure redistribution from the point of view of society, just like a profits tax. Recall that according to Eq. 3.1 the loss of profits is valued *before* taxes.

The other principal tax component relates to changes in demand for labor,

$$(t^\ell + t^w)wd\ell^s, \tag{3.6}$$

where t^ℓ is a (for simplicity proportional) tax on labor, t^w is a social security fee, w is the wage rate, and $d\ell^s$ is the change in labor supply. If supply decreases by the same magnitude as demand for labor at Dönje, then we should value labor cost at the reservation wage. That is, the cost should be valued at the marginal disutility of work effort. The result is obtained by using Eqs. 3.2 and 3.6 with $d\ell = d\ell^F$ in Eq. 3.1.

It seems unlikely that the small change in electricity generation under consideration will have any noticeable impact on aggregate demand for electricity (or other goods) or supply of labor. Therefore, in the base case we set $dx^d = d\ell^s = 0$. This issue is further addressed in the sensitivity analysis.

It should be noted that in the case of a constant labor supply we should convert the labor cost associated with $d\ell^F$ to a market value by multiplication by one plus the value added tax; see Appendix A for details. This is because the workers released at Dönje will produce goods and services that are ultimately valued at consumer prices. In other words, the marginal willingness to pay of a purchaser of a particular commodity is at least as high as its market price. In the intermediate case where $d\ell^s < d\ell^F < 0$ we evaluate a part of the labor cost at consumer price and the remaining part at the reservation wage.

3.3   Some Further Observations

A first observation relates to the fact that Eq. 3.1 is stated as a *difference* between monetary benefits and costs. This is not by chance. If the object is to achieve as high utility as possible we should undertake a particular project if and only if it increases (present value) utility, i.e. if (present value) benefits exceeds (present value) costs. From Eq. A.17 in Appendix A it is seen that our cost-benefit expression reflecting the difference between monetary benefits and costs is *proportional* to the unobservable change in utility caused by a small project. Thus if the project is socially profitable, i.e. monetary benefits exceeds monetary costs, it will increase utility. Sometimes investigators present the outcome of a cost-benefit analysis in terms of a ratio between benefits and costs or calculate the internal rate of return, i.e. the interest rate that equalizes present value benefits and present value costs. However, it is well-known that such approaches cause a ranking problem. That is, they rank projects in another order than the net present value criterion used in this book. To illustrate, using the ratio criterion a project yielding benefits equal to 5 units and costs equal to 1 unit is ranked above a project yielding benefits equal to 15 units and costs equal to 5 units (and the ranking is preserved if we use the inverse criterion, i.e. cost per unit of benefits). Still, the latter project is more profitable and hence the preferred choice according to a conventional cost-benefit analysis.[10] Similar problems are associated with the internal rate of return criterion. The reader is referred to, for example, [100] or [40] for further details.

The way we treat taxes in this book parallels the general equilibrium approach used by [64]; see, for example, their Eq. 15. An alternative to this treatment of taxes in a cost-benefit analysis is to define and estimate the *marginal cost of public funds* (MCPF). There are numerous different definitions of this concept and it is further discussed in Appendix A alongside with its cousin the *marginal excess burden of taxes*. It is important to note that a project's net present value or social profitability does not hinge on whether we treat taxes in the way suggested by Eq. 3.1 or by introducing a particular MCPF-definition. The taxes are the same and the tax revenues are affected in the same way by a project regardless of the way we group them. However, it might be more transparent to handle them explicitly rather than multiply by a factor "one point x". The reader is referred to [103] and [102] for further comparison of the way taxes are treated in this book and the marginal cost of public funds. A recent and comprehensive text on the concept of the marginal cost of public funds with many empirical applications is [36].

Let us also briefly clarify one basic assumption behind our simple cost-benefit rule. A key assumption behind Eq. 3.1 is that we compare a project to a "doing nothing" or "business as usual" alternative. This approach boils down to an

---

[10]However, since both projects are socially profitable both should be undertaken. On the other hand, if the object is to maximize benefits subject to a *binding* budget constraint, the relevant criterion is the ratio between (incremental) costs and benefits. This is the criterion used in cost-effectiveness analysis and originally derived by [179].

assumption that all other possible public sector projects are exogenous. For this reason alternatives such as investments in railroads, highways, hospitals, national parks, and air quality improvements are ruled out. This is the typical approach to project evaluations. The more ambitious procedure is the one suggested by [45]. It draws on deriving *optimal shadow prices* for different factors and commodities/projects. However, to the best of our knowledge this approach has never been implemented in real-world evaluations since the underlying general equilibrium model is virtually impossible to operationalize.

A slightly simpler way to extend the conventional cost-benefit analysis is as follows. We try to estimate a number of measures that are considered to produce the same level of utility (benefits) as the project under evaluation. The judgments might be done by natural scientists or the general public, perhaps in a choice experiment involving different projects with different attributes. In the next step the costs of each alternative are estimated. In the final step one compares the minimum-cost alternative to the cost of the considered project. In terms of Eq. 3.1 one compares its costs to the minimum cost. If the project is associated with a lower cost than the minimum cost it passes the cost-benefit test. Recall that the two alternatives, by assumption, produces the same benefits. We have not had the opportunity to use this approach in this book but he interested reader is referred to [106] for a more comprehensive presentation of this approach to social project evaluations.

Finally, the reader should note that the distributional issue has not been addressed. We simply sum across agents. This quite common approach, which strictly speaking is correct only if the welfare distribution is optimal and the project is small or marginal, could be questioned but is nevertheless employed in this book. One reason for this simplistic approach is the fact that those who are likely to gain from the considered small perturbations are ordinary people and the same holds true for those who might lose from the perturbations. We are simply unable to identify affected groups of people that could be considered very poor or wealthy or "extreme" along other dimensions.

Chapter 4
The Main Items in the Cost-Benefit Analysis

As the model presented in the previous chapter illustrates, there are a number of items to consider in a empirical CBA of a change in water use. In this chapter we will briefly discuss how to measure these items. In particular, we will address the cost items of the cost-benefit analysis while the chapter to follow is devoted to the benefit side of the two scenarios under investigation.

4.1 Calculation of the Loss of Hydroelectricity

If either of the two scenarios discussed is implemented there will be less water available for electricity generation at Dönje. It is very difficult to estimate the associated loss in profits. This is because prices vary over the day, the week, and the season. In addition the managers must account for stochastic and seasonal variations in the inflow of water into its reservoir, as well as the fact that there are many interdependent hydropower plants on the river. Thus the plant managers face a complicated dynamic profit maximization problem, even if they act as price takers. Fortum, who owns the hydropower plant at Dönje, has kindly provided us with estimates of the annual loss of revenue if water is diverted from electricity generation according to SCENARIO 1 and SCENARIO 2, respectively. However, since such detailed estimates are seldom available, in this book we will use a standard approach to obtain a rough estimate of the loss and only use Fortum's estimates in the sensitivity analysis. This standard approach does not require sensitive inputs from the managers of the hydropower plant.

The amount of power supplied to the plant's turbines by the water can be stated as follows[1]

---

[1] This equation has been taken from the Energy Systems Research Unit (ESRU) at the University of Strathclyde: http://www.esru.strath.ac.uk/EandE/Web_sites/01-02/RE_info/Hydro%20Power.htm. We assume that density of water is one ton per cubic meter.

P.-O. Johansson and B. Kriström, *The Economics of Evaluating Water Projects*, DOI 10.1007/978-3-642-27670-5_4, © Springer-Verlag Berlin Heidelberg 2012

$$kW = g \times h \times f, \tag{4.1}$$

where kW denotes kilowatts, g is acceleration of gravity (≈ 9.81 m/s^2), h denotes head of the dam or river in meters, and f denotes flow of water in m^{3}s^{-1}.

From this we can calculate the plant's revenues,[2]

$$R = ef \times p^p \times kW \times t, \tag{4.2}$$

where R denotes revenues, ef is technical efficiency of the station (turbines, generator, and so on), p^p is the producer price per kilowatt-hour (kWh), i.e. $p^s + p^b$, here assumed to be constant over the considered time period, and t is the number of hours of operation during the considered period.[3]

In this study we consider small changes in the flow of water (df) during a specified period of time and according to the two scenarios under consideration. In both scenarios the head of the dam (h) is around 33 m, the assumed change in the winter flow df^{winter} is 2.75 m^{3}s^{-1}, the number of hours during the considered winter season t^{winter} is 4,920, and t^{summer} is equal to 3,840 h. We assume that the efficiency[4] ef of the plant is 0.85. Therefore, in SCENARIO 1 the annual loss of hydroelectricity is estimated to 3.7 gigawatt-hours (GWh), i.e. 3,700 MWh, and in SCENARIO 2 the estimated loss amounts to 14.3 GWh; the reader is referred to Appendix B for details. We will next turn to a discussion how to estimate a trajectory for the electricity price p^p.

4.2   Price Forecasts

In this section we first outline the electricity and other involved markets. Then we turn to a presentation of the price trajectories that will constitute a cornerstone of our empirical cost-benefit analysis.

4.2.1   Some Basic Facts

A central part of the CBA is to construct a trajectory for the price of electricity. The point of departure is the spot price at the Nordic (Denmark, Finland, Norway and Sweden) spot market Nord Pool. Figure 4.1 illustrates the working of this market. The merit order curve yields the (by assumption constant) marginal costs of different

---

[2]We assume that no water has to be wasted (since we make no distinction between dry years with high prices and wet years with low prices).

[3]Here we are primarily interested in a uniform change in the flow of water during each hour of a particular time period.

[4]According to ([176], p. 3), the total efficiency of a hydroelectricity plant is typically 80–90%.

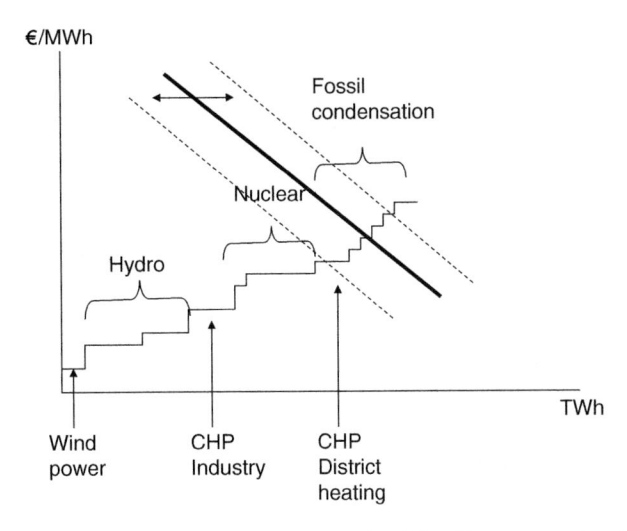

Fig. 4.1 A stylized (merit order) spot market for electricity. Plants are ordered according to variable production costs. *CHP* combined heat power

power plants. As indicated in the figure, different hydroelectricity plants, say, may have different variable operating costs. Demand for electricity shifts over the day as well as throughout season. Therefore trade on Nord Pool generates an equilibrium price for each hour.[5] Typically the marginal supplier on the spot market is a fossil-fired plant, as is also the case in Fig. 4.1. This fact means that the spot price will reflect the price of carbon emission permits. Recall that fossil-fired plants must acquire a permit for each ton of carbon dioxide they emit, which implies that their marginal costs reflects the marginal cost of permits. In recent years the price of carbon permits at the Nord Pool market has typically varied between EUR 15 and EUR 20 per EUA, with the holder of one EUA (European Union Allowance) being entitled to emit one ton of carbon dioxide or carbon equivalent greenhouse gas.[6]

Figure 4.1 assumes a perfect market in the sense that price equals marginal cost for the last unit supplied and demanded on the market, although one could conceive the possibility that low cost capacity is withhold in order to increase the equilibrium price. Some recent studies, such as [2] and [66], suggest that the Nord Pool spot market is quite competitive, so if there is market power it seems to be weak. In any case, what we focus on in the cost-benefit analysis is the value of resources released at Dönje versus the resources diverted to the expert sector in order to pay for the replacement electricity purchased on the Nord Pool market. As discussed in Sect. 3.1 the social cost of electricity production foregone is reflected by the

---

[5]Here we ignore stochastic variations in supply and demand within the hour and hence balance services provided by certain "flexible" kinds of electricity generators. There is also an "adjustment" market (Elbas) on Nord Pool where electricity can be traded after the spot market is closed.

[6]See the PDF file entitled "Marknadspriser" at www.svenskenergi.se.

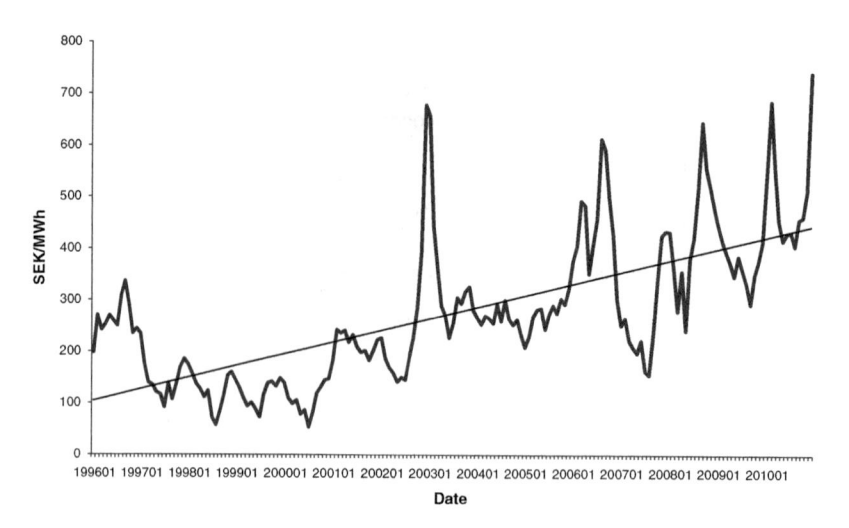

Fig. 4.2 Average monthly spot prices (system price) in SEK/MWh 1996–2010 and the linear price trend on the Nord Pool market

difference between the market price, i.e. the willingness to pay adjusted for taxes and variable transmission tariffs, and the value of the resources released at Dönje. Therefore the actual equilibrium spot price is of vital importance for this study.

In Fig. 4.2 we show how the Nord Pool spot price (system price[7]) for electricity has fluctuated from 1996 to 2010 (inclusive). The "spikes" in 2003 and 2006 are due to dry conditions, leading to a low supply of hydroelectricity, while the spikes in early and late 2010 were due to extremely cold periods and transmission problems. The figure includes the linear trend of the spot price.

There is trade between the Nordic countries and Estonia, Germany, the Netherlands, Poland and Russia. Hence the price on the Nord Pool is not independent of what happens in the electricity markets in the surrounding area.[8] For example, in the dry years of 2003 and 2006, the Nordic countries did import electricity from Germany, Poland and Russia. In particular, the spot market of the Leipzig based European Energy Exchange, EEX, is of interest in the present context.[9] This is because from the point of view of Nordic hydroelectricity producers it seems to

---

[7]There are sometimes bottlenecks in the transmission of electricity implying that the Swedish price deviates from the system price. Here we ignore such deviations since we focus on long run issues. Historical spot prices are available at Nord Pool's home page: http://www.nordpool.com/asa/. The prices shown in Fig. 4.2 are listed in Appendix C. Since inflation has been very low during the considered period we report the spot price in current terms.

[8]For a good treatment of European electricity markets the reader is referred to [156].

[9]On every exchange trading day, closed hourly auctions regarding delivery in Germany, Austria and Switzerland on the following day take place on the spot market for power. In this auction market participants can place also block bids in addition to the hourly contracts (i.e. bids for several consecutive hours). As a supplement to this, intraday trading, which permits trading of hourly

be ideal to have unlimited transmission capacity to Germany since the EEX spot price typically is higher than the Nordic one and has a different profile across the day; see below. The German merit order curve looks similar to the Nordic one but there are also important differences. In contrast to the Nordic system, Germany has little hydroelectricity and a lot of lignite-fired power plants. Germany also has pumped-storage plants for electricity generation during peak hours.[10]

There are two prices on the EEX-market that are of interest in the present context, the base load price and the peak load price, respectively. The former refers to the average price over an 24-h period on weekdays while the latter refers to the average price over the period from 8am to 8pm on weekdays. The peak load price is of most interest to flexible hydropower producers. We lack detailed price statistics from the EEX market. The average (monthly) peak load price was EUR 63 per MWh for the period 2005–2010 (inclusive[11]). However, the prices during 2009 and 2010 were heavily influenced by the ongoing worldwide economic crisis. The average over the years 2005–2008 was EUR 70 per MWh. It would be an advantage for Swedish hydroelectricity producers to be able to sell more on the EEX at peak load prices and possibly exploit the fact that EEX prices show a quite cyclical pattern over the day as is further illustrated below, although there are transmission limitations reducing the possibility of this type of export sales.

There are extensive capacity expansion plans in the Nordic countries. Denmark, Norway and Sweden are expected to massively expand wind power while Finland plans another nuclear plant; see, for example, [169]. One would expect these plans, if realized, to exert a downward pressure on prices. A massive expansion of wind power might also force hydroelectricity plants to reoptimize their supply pattern over the year since wind power cannot be adjusted in the same way as hydropower.

Germany, like several of the Nordic countries, plans a huge expansion of its wind power. Figure 4.3 illustrates how recent expansions have affected the German spot price according to [178, p. 9]. In particular, during peak hours wind power reduces the spot price. A further expansion would strengthen this pattern. On the other hand, if Germany shuts down its nuclear power plants, one would expect upward pressure on prices. It should be noted that Germany's expansion of wind power is planned to take place in the northern parts of the country, which would aggravate Swedish plans to export hydroelectricity to Germany since transmission capacity is limited.

---

deliveries on the same day around the clock up to 75 min prior to the beginning of delivery, was launched, at first, for Germany in September 2006.

[10] A pumped-storage plant (typically) works as follows. At night, when electric demand is low, the plant's reversible turbines pump water (uphill) to a reservoir. During the day, when electric demand is high, the reservoir releases water through the turbines.

[11] The average (monthly) peak load price was EUR 63 per MWh in 2005, EUR 73 in 2006, EUR 56 in 2007, EUR 88 in 2008, EUR 51 in 2009, and EUR 55 in 2010. This yields an average of EUR 63 (EUR 70) for the period 2005–2010 (2005–2008). Since the credit crunch affected the last 2 years of the period we set the expected normal price at EUR 70 (in 2010 prices). Source: Annual reports of the German company RWE AG (http://www.rwe.com/web/cms/en/108460/rwe/search/?q=annual+reports).

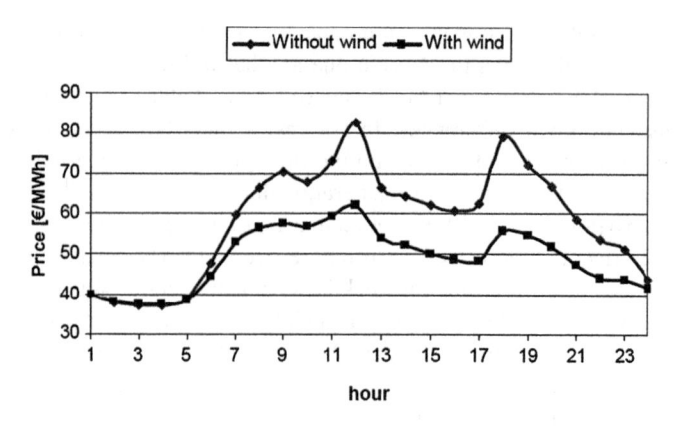

Fig. 4.3 Average hourly German spot prices 2006–June 2008 with and without wind power (Source: Fig. 4.2 in [178])

Figure 4.3 also illustrates the cyclical price pattern over the day in Germany, where pumped-storage plants are currently used to produce electricity during peak hours. The ideal scenario for Swedish hydropower producers would be to be able to crowd-out these pumped-storage plants during peak hours and likely receive a considerably higher price than on the Nordic market.

In closing let us mention recent plans to further connect different European submarkets. There is a memorandum of understanding between the national grids of the three Baltic states and their Swedish counterpart to build an underwater power cable between Lithuania and Sweden. When this power cable, which will be partly financed by the European Union, is opened (probably in 2016) the Baltic states will be fully integrated into the Nordic electricity market. Statnett (the national grid of Norway) and a subsidiary of UK's National Grid have signed a contract to explore the prospects of an underwater power connection between the two countries. The reason is that the UK plans to invest heavily in wind farms and therefore needs trading partners to balance a system with highly variable electricity generation. Norway will supply hydropower electricity on calm days while the UK will supply Norway, and hence the Nord Pool, market with wind power electricity on windy days. The power companies awarded contracts to build 6,400 wind turbines off the British coast, which are expected to cost around SEK 1,200 bn and to be completed by 2020, have warned that "they will need a 'super-grid' connected to Europe to guarantee a steady power supply" (The Daily Telegraph, January 9, 2010, p. B2). A cable connection to the UK is one of several projects Statnett plans between Norway and Europe. Other connections are cables to Denmark and directly to continental Europe. It remains to be seen how these and other massive expansion plans of European wind farms and international power connections will change Nord Pool equilibrium prices over future days, weeks and seasons.

4.2.2 Two Price Trajectories

The price series in Sect. 4.2.1 are historic ones. In the CBA we need forecasts for future prices or price trajectories. Ideally, we would like to estimate systems of interdependent price equations, either reduced-form or structural. For example, one would like to specify an equation for the Nordic spot price, an equation for the German spot price, an equation for the carbon permit price, and possibly equations for oil, coal, and gas prices. Once such a system has been estimated, one would use the coefficients to forecast the Nordic spot price. There are of course also other econometrical approaches, for example, time-varying regression models, which can be used to test for structural changes in electricity markets, models with Markov regime switching, which can be used to test how prices react to temporal irregularities in markets, generalized least squares (GLS) structural models, that can be used to test if residual unconditional variance respond to fundamentals, and Garch models, which can be used to shed light on whether residual conditional volatility react to past volatility and shocks; see [26] for an interesting discussion along these line.

For this book such econometric approaches were not possible to implement although we plan to explore them as part of a future project. Instead, we will use a simple ad hoc approach or bold guesses in order to generate two different trajectories for the electricity price used to estimate the loss of revenues for the power plant at Dönje.

Before turning to that discussion let us first sum up a number of factors that might affect the Nordic spot price. Firstly, one would expect reductions over and above those already announced for the number of emission permits due to the strong concerns for the global climate. Such reductions would put upward pressure on the spot price as long as the marginal plant on the Nordic spot market is fossil-fired. Secondly, if fast-growing nations like China and India continue to grow at a high rate one would expect oil, coal and gas prices to continue to rise. This would also put upward pressure on Nordic spot prices as long as the marginal plant is fossil-fired. Thirdly, there are strong forces supporting a single European market for electricity (subject to technological an economic limitations). In practice this means that there is likely to be a pressure to improve transmission lines between different European markets. If transmission capacity between the Nord Pool market and the German EEX market is strengthened prices will probably be determined on the much larger German market. Fourthly, capacity expansion, for example a massive wind power expansion, works in the opposite direction. Due to all these uncertainties we will introduce two distinct trajectories for the Nordic spot price.

The first price trajectory is a conservative one which is aimed at providing a reasonable *lower bound* for the loss of revenues of the Dönje plant. According to this scenario, the price received by Dönje "today" is equal to our estimate of Nord Pool's (system) spot price for an average or "normal" year with respect to precipitation.

This price, assumed to be constant over time, is set to SEK[12] 350 per MWh (about EUR 35 per MWh) and corresponds roughly to the average spot price during the period 2003–2010; the average is SEK 361 per MWh. This forecast is consistent with the idea of mean-reversion, which is the tendency for a stochastic process to return over time to a long-run average value. The reader is referred to [107] for further discussion of the concept of mean-reversion in the context of electricity markets.

The second price trajectory is supposed to provide a reasonable *upper bound* for the loss of revenues for the Dönje plant. The best scenario of Swedish hydropower proponents is arguably a "merger" of Nord Pool and the German EEX market. In this scenario Swedish hydropower is sold at the German peak load price from 2030 and on. We assume that the initial average German peak load price is SEK 700 per MWh (about EUR 70 per MWh), which corresponds to the average price during the period 2005–2008, and that the average price stays constant at this level. As in the first scenario, the initial price received by the Dönje plant is assumed to be equal the "normal" annual spot price on the Nord Pool market, i.e. SEK 350 per MWh, and then linearly approaches the German peak load price in such a way that they become equal in 2030. This corresponds to an annual real price increase of 5% during the period 2010–2030 so that the real spot price reaches SEK 700 per MWh in 2030 and then stays constant.

If we use the simplest extrapolation model, i.e. the linear trend model, to forecast future Nordic spot prices, we obtain a price of almost SEK 900 per MWh for the year 2030.[13] The time series is the one shown in Fig. 4.2. Even if such simple extrapolation methods can be useful as a way of quickly formulating a forecast, they typically have little accuracy. In later research we will estimate alternative and more realistic econometric models. However, in this book we stick to the aforementioned estimates of lower and upper bounds for the spot price trajectory.

It might be argued that the first price trajectory represents "business as usual" in the sense that there are no major changes in the market. The second trajectory accounts for a possible major long-run structural change in market conditions. In our cost-benefit analyses we will draw on point estimates of the different items. With respect to the price trajectories we assume arbitrarily that the spot price at each point in time is an iid uniform random variable over the interval (p_t^{slb}, p_t^{sub}), where p_t^{slb} (p_t^{sub}) denotes the lower bound (upper bound) estimate of the spot price at time t. This is the simplest possible continuous random process and means that the point estimate of the spot price at time t is equal to $(p_t^{slb} + p_t^{sub})/2$. This means that the initial assumed spot price is SEK 350 per MWh and that the price increases linearly to reach SEK 630 per MWh after 20 years and then stays constant in real

---

[12]A Swedish krona (SEK) is here assumed to be worth about EUR 0.1 in the long run, i.e. EUR 1 is equal to SEK 10.

[13]The linear trend model is $p_t = 102.9 + 1.9t$, with t-statistic equal to 6.7 and 12.9, respectively, $R^2 = 0.48$, adjusted $R^2 = 0.48$, and the low Durbin-Watson (DW) statistic, DW $= 0.31$, indicates that serial correlation is present in the estimated residuals.

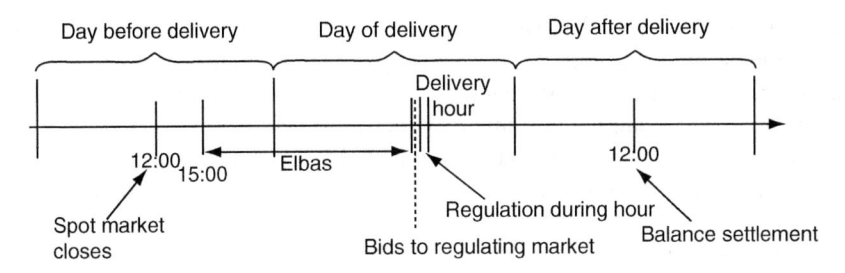

Fig. 4.4 Time frame for the working of the Nordic electricity market (Source: Fig. 2.5 in Olsson [147])

terms. This trajectory yields the same present value revenue loss as if we use what might be termed a *certainty equivalent price* of just above SEK 480 per MWh,[14] i.e. the same present value revenue loss as if the price is constant over time and equal to SEK 480 per MWh. In the sensitivity analysis we will show what range of present value losses of revenue our spot price bounds generate. As a comparison it might also be noted that the Swedish Energy Agency, which is a government agency for national energy policy issues, in a recent long-run forecast (see [166]), estimates the spot price (or rather Swedish system price) to be SEK 500–520 per MWh over the next 20 years in 2010 prices. In contrast to one of our price trajectories this last forecast seemingly ignores any major changes in international transmission lines.

A hydropower plant can typically quickly adjust its production level and therefore provide balance services that are valuable; see, for example, [147] for a detailed discussion. For example, demand might suddenly fall or a plant break down. For the stability of the system it is then important to have plants that very quickly can adjust so as to stabilize – primary regulation – and restore – secondary regulation – grid frequency (in the Nordic countries 50 Hz). These are denoted regulation during hour in Fig. 4.4.

There is also a need for adjusting power. This is because bids for the spot market can be submitted only until 12:00 the day before delivery and at 15:00 the prices and quantities for each hour of the next day are released. Thereafter and up to 1 h before delivery adjustment electricity can be traded on the Elbas market. However, the prices recorded at Nord Pool's Elbas market seems to indicate that the extra value of such services is currently low; expressed as SEK per MWh hydropower, the current value seems to be less than SEK 10 per MWh hydroelectricity.[15] Moreover, given that we consider uniform changes in the water flow, i.e. a fixed number of m^3/s, Dönje's possibility to supply regulating power is not seriously harmed (according to Mr. Kent Pettersson, Fortum). This fact motivates ignoring such services in this

---

[14]Unless otherwise stated we assume that society is risk neutral. The certainty equivalent price is defined in Eq. B.3.

[15]Only a small fraction of the total output of hydroelectricity is traded on the Elbas market. Therefore, the "bonus" expressed as SEK per MWh of hydropower produced is small.

study. This is not to say that we dismiss the role of hydropower as a regulating source. On the contrary, since this role is likely to become increasingly important in the future, for example as more and more (stochastic) wind power is added to the system. We plan to return to this important topic in future research.

Finally, it should be stressed that we ignore the fact that a hydropower plant can store water in order to produce electricity at times when prices are high. However, it is not entirely obvious whether our approach causes an overestimation or an underestimation of the loss in revenue of the considered plant. The exact outcome depends on whether it is assumed that our price trajectories represent peak prices during the winter season or prices during the summer season. Our approach is simply too rough to provide a simple and clear answer.

4.3 Choice of Discount Rate

There is a huge literature on how to define and estimate a social discount rate. There is certainly no universal consensus with respect to what a social discount rate reflects, its magnitude or even sign. A good overview of different approaches is found in [175]. We will not attempt to summarize the different approaches and views here. We just note that the UK uses a base rate of 3.5% (including 1% reflecting catastrophe risk), Germany's rate is 3% while France uses a rate of 4%; see [50]. Reference [51] argues for a standard benchmark European discount rate of around 3–4% based on social time preference. The European Commission's Directorate General Regional Policy has suggested a 3.5% social discount rate for most member states including Sweden when evaluating infrastructure investments; see the Commission's guide to cost-benefit analysis; [50].

Both [175] and the Commission's [50] CBA guide provide estimates of the Swedish social discount rate. The model used by [175] is the one suggested by [52]:

$$r = (1 + \dot{p})^{1-\alpha}(1 + \dot{y})^{\sigma}(1 + i) - 1, \qquad (4.3)$$

where r is the discount rate, $\dot{p}$ is population growth, α gives the weight of population size on social utility, $\dot{y}$ is per capita income growth, σ is the coefficient of relative risk aversion, and i is the pure time preference. According to this model one arrives at discount rates of 2.9–3.4% if the annual growth rate of per capita income is about 1.8% as was the case for the period 1970–2008; the higher (lower) rate is obtained if the coefficient of relative risk aversion (σ) is 1.26 (1 as in their base case[16]). It might be noted that the result that $\sigma = 1.26$ was obtained as a best estimate by [123]

---

[16]Using Eq. 8 and Table 7 in [175] $r = 100 \times (1.0024^{0.5} \times 1.018^{\sigma} \times 1.01 - 1)$, where σ is either 1.26 or 1. The estimate is based on GDP per capita rather than income per capita (which was not available for the considered time period). See Statistics Sweden: "National Accounts, quarterly and preliminary annual calculations".

in a cross-sectional study covering over 50 countries and time periods; the reported estimates are seemingly very robust and vary from roughly 1.2 to 1.35.

The model of the European Commission (it's Directorate General Regional Policy, see [50]), is as follows:

$$r = \dot{c} \times em + i, \tag{4.4}$$

where $\dot{c}$ is the growth rate of consumption (i.e. $(dc/dt)/c$), em is the elasticity of marginal utility with respect to consumption,[17] and i is the pure time preference. Using this model and assuming that the annual growth rate of Swedish consumption is 1.7%, one arrives at a discount rate of 3.1%.[18] It should be mentioned that the social discount rate defined in Eq. 4.4 is a general equilibrium rate corresponding to the classic infinite horizon Ramsey model that is analyzed in all advanced macroeconomics courses and nowadays in an augmented version (as in Eqs. 2.3–2.6 in Chap. 2) in a typical advanced course in environmental economics. Maximizing a present value Hamiltonian yields first-order conditions that can be rearranged to yield the rate as defined in Eq. 4.4 which then equals the marginal product of real capital. The reader is referred to Eq. 7' on p. 40 in [18], with the rate of population growth set equal to zero, or Eqs. 2.8 and 2.10 in [7].

Even with moderate constant discount rates, large future damages have almost no effect on current decisions. For this reason it has become quite common to argue in favor of hyperbolic discounting; see, for example, [38] and [110]. Such an approach results in a discount rate that is decreasing over time. Thus future generations are attributed lower discount rates than current ones. In the present context this approach of hyperbolic discounting does not seem to be relevant. The reason is the fact that the decision to change the water use is reversible in both directions at any point in time. At least from a theoretical point of view this means that any generation can sell or buy water use rights. Therefore, we will assume a constant discount rate. It does not seem implausible that there is a consensus emerging within EU that infrastructure project should be assessed using a discount rate of around 3–4%; the rates mentioned above suggest such a possibility. It would greatly simplify comparisons of different infrastructure investment evaluations if they used the same discount rate in their base case evaluations. This does not prevent authors from strongly arguing for and applying different discount rates in their sensitivity

---

[17]From a technical point of view em equals a coefficient of relative risk aversion. However, in [50] it seems as if em and $\dot{c}$ refer to public rather than private consumption.

[18]Using data stated in Table B.2 on p. 207 in [50] but with $\dot{c} = 1.7$ (instead of $\dot{c} = 2.5$) one obtains $r = 1.7 \times 1.2 + 1.1 = 3.1$. The annual growth in real (public as well as private) consumption was around 1.7% during the period 1970–2006 (and no data are currently available beyond 2006). See Statistics Sweden: "Detailed annual national accounts 1993–2006, some series from 1950 to 1980 (Corr. 2009–02–24)".

analysis. In any case, we will set the social discount rate equal to 3%; this rate is in accordance with the Swedish estimates presented above.

In the sensitivity analysis the discount rate will halved and doubled, i.e. changed to 1.5% and 6%, respectively, in order to illustrate the sensitivity of the results for the choice of discount rate. It might be noted that the lower rate is close to the one proposed by [167] in his famous "Stern Review" of the costs of climate change while the higher rate is the one proposed by [182], one of Stern's more prominent critics among economists. However, it should be stressed that they evaluate large irreversible changes that might be associated with catastrophic consequences, i.e. "projects" that are quite different from the marginal ones under consideration here. In other words, it is far from self-evident that the rates used in evaluating global climate change are relevant for our scenarios.

Another issue relating to discounting is how to estimate the minimum compensation that Fortum requires in order to be willing to undertake the considered changes in electricity generation at Dönje. If the firm uses another discount rate than the one used in a CBA, then we have two different present value amounts, depending on what discount rate is used. However, if Fortum knows the social discount rate, it would not use a higher rate than the social one when computing its present value compensation claim. The reason is simply the fact that the present value is a decreasing function of the discount rate, all other things constant. If the social discount rate is not known to Fortum, the outcome would probably depend on the parties bargaining power, and it is not obvious who would be on the opposite side of the negotiation table, for example, the local municipality or the national government. Alternatively, if Fortum receives annual amounts corresponding to the annual profit loss, the relevant present value for the CBA is calculated using the social discount rate. In this book we will adopt this approach and discount the loss of profits at 3% in the base case. However, in Sect. 6.3 we return to a further discussion of this important issue.

4.4   Hydroelectricity Production Costs

Hydropower plants have low or zero variable costs. Therefore, Fortum will probably save very small amounts in the short run if production at Dönje is reduced. For example, [78] in their detailed analysis of the cost of building different types of power plants claim that the variable cost of new hydropower is virtually zero. The same claim is made by the large Swedish power company Vattenfall; see [177, p. 18]. However, most merit order curves indicate that *existing* plants have strictly positive marginal costs. Therefore, in our base case we assume arbitrarily that the marginal cost is SEK 30 per MWh for the *existing* plant.[19] Sooner or later

---

[19] According to the cost-benefit rules derived in Appendix A and summarized in Chap. 2, in the base case the marginal cost should be converted to market price by multiplication by one plus the VAT rate. Therefore, our marginal cost is assumed to include VAT.

the turbines and other equipment must be replaced by new ones; it is arbitrarily assumed that the current one is operated for an additional 40 years but in the sensitivity analysis it is shown how the outcome is altered if the remaining life span is shortened. In any case, it seems unlikely that the kind of marginal change in water use we consider will affect the magnitude of the future investment cost. Therefore, in the base case this cost is ignored.

However, in the sensitivity analysis we will assume that the magnitude of the investment cost is proportional to water use. The unit or average fixed cost of the investment is taken from [78]. They estimate the cost to about SEK 170 per MWh (net of taxes) if the social discount rate is 3% and the plant has an economic lifetime of 40 years.[20] The present value cost today per MWh of an investment undertaken at time τ and with an economic life of Υ years is calculated as,

$$I_{MWh}^{\tau}(r) = 170 \int_{\tau}^{\tau+\Upsilon} \frac{\sum_{t=0}^{39}(1+0.03)^{-t}}{\sum_{t=0}^{39}(1+r)^{-t}} e^{-rs} ds, \tag{4.5}$$

where $I_{MWh}^{\tau}(r)$ denotes the present value investment cost per MWh today of an investment undertaken at time τ and having an economic life of Υ years, and r is equal to 3% in the base case. Multiplying by the number of MWh lost due to the considered changes yields the cost savings due to a lower future investment cost at time τ. Referring to [78] we ignore any variable cost of running the new plant and the scrap value of the investment is assumed to be zero. Hence letting Υ approach infinity in Eq. 4.5 provides us with an estimate of the present value of the total cost savings.

4.5 WTP/WTA for Environmental and Recreational Effects

Each considered reregulation or scenario affects a number of externalities. These need to be added before arriving at a complete cost-benefit analysis of our proposed scenarios. To recap, net benefits associated with a scenario equal $WTP^{RB} + WTP^{B} - WTA^{E}$, where the items in turn are related to the willingness-to-pay for obtaining an improved downstream river basin, the willingness-to-pay for a smoother downstream flow of water and the willingness-to-accept compensation in lieu of increased emissions of climate and other gases (sulphur, nitrogen, and so on). The Contingent Valuation study captures the first of these components,

---

[20]Reference [78] present results for discount rates equal to 6% (about SEK 250 per MWh) and 12% (about SEK 430 per MWh). We have used these results to arrive at a cost of SEK 170 when the social discount rate is equal to 3%.

while the two other await detailed empirical studies. Nevertheless, we will indicate approaches to pin down the values for these two terms. Our focal point is WTP^{RB} and in Chap. 5 we briefly describe the Contingent Valuation study that was used to estimate it, as well as possible empirical approaches for the estimation of the other non-priced items.

Chapter 5
Towards Empirical Measurement of Externalities

The re-regulations considered in this book effect the externalities caused by hydropower production. This chapter details the approaches used to estimate the value of non-market goods and services attributable to the reregulations. We begin by a short overview of different values that a natural resource such as river might provide. We then turn to a brief discussion of different methods that can be used to estimate the willingness-to-pay for the services provided by our projects and the reasons for undertaking a contingent valuation experiment in order to estimate the main non-priced services provided by our projects. We then describe how the ecological assessments were carried out and turn the our contingent valuation study. This study is comprised of the focus group analysis we used as an integral part of designing the survey instrument; the results of this analysis are outlined briefly. We then turn to pertinent details of the pilot study, including the estimates of the three WTP components that are a central ingredient of our cost-benefit analysis.

5.1 Use Values and Non-Use Values

There is an obvious reason why a commodity is valued, whether it is produced or a service provided by a natural resource or a species. It simply provides *use* values, i.e. it is an argument in individual utility functions. This is obviously the case for commodities like food, drinks, television programs, movies, and so on, but the same also holds for environmental commodities. For example, a river provides different recreational services like fishing, canoeing, and kayaking. In addition the river and its surroundings might provide scenic beauty and other aesthetic values. These values too are a type of consumption and hence are arguments in individual utility functions. It might be added that it is a completely different question wether a service is or can be priced in a market. All kinds of commodities (i.e. goods and services) from pure public ones to purely private ones that are consumed, generate what we here term use values; the reader is referred to, for example, [20] or [140] for definitions of different types of private and public goods.

P.-O. Johansson and B. Kriström, *The Economics of Evaluating Water Projects*,
DOI 10.1007/978-3-642-27670-5_5, © Springer-Verlag Berlin Heidelberg 2012

A resource or a service might be valued even if it is not consumed. Such values are often referred to as *non-use* values but sometimes they are labeled *passive-use* values or *intrinsic* values. Non-use values in a broad and imprecise sense can be traced back to conservation movements in different countries of the late 1800s and early 1900s. As noted in Chap. 2, economists started to look at the concept in the 1960s. In his seminal article [119, p. 781], observes that "There are many persons who obtain satisfaction from the mere knowledge that part of the wilderness of North America remains, even though they would be appalled by the prospect of being exposed to it." In terms of a simple utility function one could interpret Krutilla's observation as meaning that $U = U(c, NUV)$, where c is private consumption or use values, and NUV denotes non-use values. In other words, if the existence of a resource or species generates utility, it has a value that should be accounted for in a cost-benefit analysis.

There are several different categories of non-use values. Above we mention *existence* values, where the survival of a species or the preservation of a resource is attributed value. To illustrate, one might attribute value to the preservation of a species, e.g. blue whales, even though one will never see ("consume") one. A person might also positively value the *option* to consume a resource sometimes in the future. This is referred to as an option value, although the term option value has a broader meaning, as is obvious from Chap. 2. Still another category of non-use values might be labeled *altruistic* values. One might care about the possibility of others to consume a resource even if one does not consume it oneself. It might be the current generation or future generations, sometimes referred to as bequest motives, that one cares for. In turn, one might be a *pure* altruist in the sense that one respects the preferences of others. The utility function of individual h might look like $U_h = U_h[c_h, \mathbf{U}_{\neq h}]$, where $\mathbf{U}_{\neq h}$ is a vector of utility functions, except the one of h. An illustration is provided by our assumption that Swedes care about the impact on foreigners of emissions caused by replacement power produced by coal-fired Danish plants when the electricity generation at Dönje is reduced although our actual estimate discussed in Sect. 5.6 more resembles what below is termed paternalistic altruism. Altruism might also be *paternalistic*. For example, one might be concerned with how a project affects income distribution in society (income-focused altruism) or how it affects health of different groups of individuals (safety-focused altruism). Finally, the pure act of giving is sometimes assumed to provide utility. This is referred to as *impure* altruism or warm-glove giving. For deeper and more extensive treatments of use values as well as non-use values the reader is referred to any textbook in environmental economics.

In the present study we focus on use and non-use values experienced by the local community and altruistic concerns with respect to emissions caused by replacement power. The reason is the fact that the small perturbations that are considered in this study are very unlikely to involve any use or non-use values outside the local community. A possibly fruitful alternative would be to estimate the minimum cost of achieving the same benefits as those provided by our two scenarios by *other* measures, possibly somewhere else in the country. This approach was briefly discussed in Chap. 2 but it was developed too late to be implemented in this study.

5.2 Empirical Approaches for the Valuation of Environmental Benefits

Ideally there are perfect markets for the good(s) under evaluation. Then the equilibrium price informs about the marginal willingness-to-pay for the good. However for the environmental goods and services that are relevant in the present context there are no markets. Therefore other approaches are needed. There are basically two broad categories of approaches. One approach draws on prices in markets related to the good under evaluation. For example the market price of a home might reflect the value of local environmental attributes. Using econometric techniques one can estimate the value of or willingness-to-pay for the environmental attributes. Such methods are termed *revealed preference* methods or surrogate market methods. These methods can be used to estimate use values but cannot capture non-use values. The last property is due to the fact that non-use values leave no "fingerprints" in markets. The second principal approach collects information through surveys, in-person interviews, telephone interviews, mail surveys, or interactive computer programs, hereafter referred to as web-surveys. For example, a sample of people might be asked about their willingness to pay for a set of environmental attributes. This is referred to as the contingent valuation method and the first academic application was probably the early 1960s work by [39]; the seminal work on this method is [138]. Alternatively the sample is confronted with attributes of the scenario and these attributes are varied allowing respondents to accept or reject a particular change in the attributes. This approach for assessing the marginal willingness-to-pay, developed by researchers in marketing in the 1960s, is often termed choice experiments although other names such as conjoint analysis are quite frequent in the literature. These survey-based methods are known as *stated preference* or non-market methods.[1] A recent non-technical review and assessment of stated preference and revealed preference methods is found in [62] and [61] provides a full coverage of available approaches. A good treatment of stated preference methods is [9]. A recent high-quality text on environmental economics, including the methods under discussion, is [115].

There are several different revealed preference techniques. The hedonic pricing method assumes that the price of a marketed good is related to its characteristics of the services it provides. For example, the price of a property reflects its location, its size in m^2, number of rooms, number of bathrooms, closeness to shops, schools and work places, environmental attributes, and so on. Therefore, one can estimate how price or willingness-to-pay is changed when a characteristic of the property is changed. Similarly, the hedonic travel cost method derives implicit prices of recreational attributes by regressing travel costs on a bundle of attributes associated

---

[1] There are other methods than those covered here, for example, benefits transfer and dose-response functions. The reader is referred to a textbook in environmental economics for coverage of such methods but see also below.

with the destination sites.[2] The more traditional zonal travel cost method basically estimates number of trips to a site as a function of travel costs, income and various socio-economic characteristics. The area under the estimated demand curve above the travel cost provides an estimate of the consumer surplus associated with trips to the recreational site in question. We might add that the origins of the travel cost method dates back to a letter from the ingenious economist and statistician Harald Hotelling to the Director of the U.S. National Park Service in 1947; see [88]. The Director had asked a number of economists how to value a national park. In his letter Hotelling noted that people travel considerable distances and hence incur travel costs. He goes on to describe how these cost figures can be used to estimate the aggregate consumer surplus in the way suggested above.

Turning to stated preference approaches, choice experiments are a technique for establishing the relative importance of different attributes in the provision of a good. It assumes that any good can be defined as a combination of levels of a given set of attributes. The utility derived from the good is thereby determined by the utility of each of the attributes. There are five stages in a choice experiment. The first step is to establish the relevant attributes out of possibly a huge amount of attributes. The second step is to assign levels to the attributes, for example, an attribute might be assigned three levels (good air quality, intermediate, bad). In the third step one decides on which scenarios to present. If each attribute has three levels and there are four attributes there are 81 possible profiles or scenarios. The number must be restricted in order for the respondents to survive the experiment. The fourth step is the empirical survey generating a data set which is used to establish preferences. Respondents face a set of characteristics and choose an alternative. Then attributes are changed and new choices are made. For example, air quality is improved from fair to good but the cost increases from SEK 100 to SEK 150. The respondent has to decide whether the initial or the final situation is better. These choices are then subjected to regression analysis in order to estimate the "shadow prices" of different attributes. In the final step utility is estimated. For book length accounts of how to use this method in valuation studies, see e.g. [13] and [109].

As noted, contingent valuation is a survey technique that aims at a more direct assessment of the willingness to pay. It presents the respondent with a description of the project or scenario under evaluation. There is a choice question from which values are inferred. In the simplest case the respondent is simply asked his maximal WTP for the considered project. This elicitation method is termed an open-ended valuation question. Alternatively, the respondent faces a closed-ended or binary choice where he or she has to accept or reject to pay a specified amount of money for the project. This amount is varied across subsamples of respondents and econometric techniques are used to estimate the mean WTP for the sample. Many variations of these two basic valuation mechanisms exist and a new variation involving self-selected intervals for one's uncertain WTP is introduced in this study.

---

[2]It has been argued that the random utility model (RUM) is superior to the hedonic travel cost model. Reference [150] provides a detailed comparison of the models.

The probably single most serious problem with the contingent valuation approach, shared with other survey-based methods, is that choices are not binding since no real payments are made. Therefore it is an open question whether answers correspond to a real-world willingness-to-pay. Considerations of space prohibits us from digging into all the possible bias problems that are associated with different valuation methods. The reader interested in reading more is advised to consult a textbook in environmental economics.[3]

Instead we turn to a brief discussion of how our three "externality"components could be measured. We begin with WTP^{RB}, the willingness-to-pay for obtaining an improved downstream river basin. Recreational uses, such as fishing, canoeing and kayaking can be estimated using zonal or hedonic travel cost methods or random utility models. That might also be true for aesthetical improvements. Still there might be a WTP among local residents who do not necessarily derive use values themselves from the scenarios but value that others enjoy an improved recreational environment, i.e. there might be altruistic concerns. If the projects secure the survival of species there might also be a WTP related to existence values, or passive use values as discussed in Chap. 2. Such non-use values require stated preference approaches such as contingent valuation or choice experiments. Alternatively one uses stated preference methods to obtain estimates of the *total* WTP of a sample from the population of affected individuals/households instead of using different methods for different components of total WTP.

The willingness-to-pay for a smoother downstream flow, WTP^B, might be reflected in prices of houses situated close to the river since an attribute is changed. However, many of those who benefit from a smoother flow are not the "on-site" home-owners. Those who benefit in this sense visit the river. One characteristic determining the number of trips to the river might be how smooth the flow is. Thus the hedonic travel cost method might be used to shed some light on the willingness-to-pay for a smoother flow. Alternatively, stated preference methods might be used. However, as is further explained in Sect. 5.5 the physical impact on the smoothness of the flow is so small that the best approximation is that $WTP^B \approx 0$.

The willingness-to-accept compensation in lieu of increased emissions abroad, $-WTA^E$, is a non-use value. In this case the most obvious valuation candidate is a stated preference method. If one is unable to collect this kind of information, most likely due to financial constraints, one could try to collect information on the shadow prices of additional emissions of different particulates and then aggregate so as to obtain an estimate of the total "value" of a change in emissions.[4] The problem

---

[3]For more specialized treatments on how to deal with the many practical issues using direct approaches, see e.g. [8] and [9]. Books that focuses more on the econometric aspects include [71] and [130]. A perspective on the enormous research output in valuation and its relatively scant influence on policy is given by [1].

[4]Such approaches involve benefits transfer, where, for example, values have been estimated for one location and are "transferred" to another location, possibly in another country, and the use of dose-response functions that provide estimates of the relationship between the dose or size of a pollutant and its impact on morbidity or mortality.

of course is to obtain the relevant shadow prices for different particulates. This is further clarified in Sect. 5.6. There are also web-based tools that can be used to obtain estimates of the cost of different emissions changes. These tools are both easy to use and free of charge. We will use such a web-based tool, EcoSenseLe, to estimate the magnitude of $-WTA^E$. This approach is further described in Sect. 5.6.

Let us return to the single most important benefit of the considered proposals, WTP^{RB}. In order to obtain a single WTP-measure for each affected individual we have chosen to undertake a contingent valuation experiment. The basic reason for ruling out choice experiments is that such experiments do not provide a direct measure of the willingness to pay for a particular re-regulation or scenario. To motivate this choice further we need to explain some basic features of these two approaches.

Choice experiments, and versions of contingent valuation that use closed-ended valuation questions, are based upon estimating parameters of the utility function. The dominating approach in this endeavor is the Random Utility Maximing Model (RUM) due to Daniel McFadden. The exposition of RUM below is based on [118].

Suppose a decision maker is about to select one of J different alternatives in a choice set C and let z_j denote a vector of attributes describing the jth alternative ($j = 1, \ldots, J$). The utility obtained from the jth alternative can be expressed as

$$U(z_j, m) = V(z_j, y) + e_j, \tag{5.1}$$

where $V(x_j, m)$ is non-stochastic and is the utility derived by the "representative" decision maker, m denotes income, and e_j is a stochastic individual specific component.

If the decision maker is rational and maximizes utility, the jth alternative is chosen if and only if (iff)

$$U(z_j, y) > U(z_k, y) \text{ for all } k \neq j, k \in C. \tag{5.2}$$

In the standard binary response model [77], $J = 2$ and the individual is typically to reject or accept a cost for an improvement. Let z_1 denote the base case offered at zero cost ($b_1 = 0$) and let z_2 be the scenario offered at the cost b_2. The cost (or bid) is accepted iff

$$V(z_2, y - b_2) + e_2 > V(z_1, y - b_1) + e_1 \tag{5.3}$$

$$\Longleftrightarrow$$

$$(e_2 - e_1) > -(V(z_2, y - b_2) - V(z_1, y - b_1)). \tag{5.4}$$

By assuming a distribution for $e_1 - e_2$ and a particular utility function, the distribution for willingness-to-pay can be derived (see [77]). When $J > 2$ the respondent is to choose between more than two alternatives this necessarily complicates the model. Most applications follow McFadden's ingenious derivation for the general case, which essentially involves making a specific assumption about the distribution of the error terms. In particular, if we assume that the the error terms are i.i.d.

extreme value distributed terms one obtains the Multinomial logit model. This model has been generalized in various directions, such as the nested logit model and random parameters logit models.

In general, the individual specific error component is supposed to contain effects of unobserved attributes and taste variation among decision makers. Some authors interpret the error terms as representing a difference between what the researcher and the individual knows about the utility function. Components that cannot be observed or modeled are thus lumped into the error terms. Reference [118] critiques this assumption and, essentially, proposes a decomposition of the error terms, in which one term depend on conditions within the scenario, while another part is assumed independent of the scenario. This decomposition highlights a significant assumption in the standard applications of choice experiments, namely that the state of the world is completely specified by the scenario. If the utility of the scenario depends on something that is not described, this becomes an issue. In a contingent valuation study using open-ended valuation questions, the situation is quite different. There is one new state of the world and no further assumptions are introduced, we simply ask the individual about his maximum willingness to pay via an open-ended question. In this way, the individual takes care of adding any additional aspect of the scenario and just report his valuation. In a choice experiment we need to estimate this quantity indirectly and add a number of assumptions.

Thus, we need to assume particular functional forms for the utility function and the error distribution. Furthermore, the optimal design of a choice experiment boils down to making further assumptions not needed if contingent valuation is used. From an economist's point of view, one of the more problematic assumptions underlying choice experiments is that it typically assumes a linear utility function. Such an extreme assumption is seldom introduced in other parts of the economics discipline. In short, a choice experiment is based on many assumptions that are not needed in a contingent valuation experiment.

All these issues aside, even if a given choice experiment provides an estimate of the marginal willingness-to-pay for a change in an attribute, it is difficult to relate it to a particular project. It is in a sense a quite abstract willingness-to-pay measure. It should be stressed that choice experiments are extremely useful if the investigator is interested in understanding the market for a product (and we can give "the market" a very wide interpretation to allow for public goods). In a typical cost-benefit analysis, however, our view is that choice experiments are less useful for reasons explained above.

Thus, we have chosen contingent valuation as our measurement approach and in the section to follow we provide a more detailed presentation of the web-based contingent valuation experiment. Within this approach, a number of choices needs to be made, a key issue being the elicitation mechanism. As noted above, we chose self-selected intervals, where the individual reports his valuation by reporting an interval of choice (this includes, at least theoretically, a point). In what follows, we will motivate this choice and briefly explain the various statistical methods available to confront such intervals; while the data are very similar to what one encounters in survival analysis, it turns out that there are subtle differences and some additional thought is necessary. We will begin unwrapping our study, however, by providing additional details on the natural science investigations that underlie our project.

5.3 The Impact of Changed Water Flow at the Hydropower Plant

The natural science team of the research project carried out detailed experiments at the power plant, in order to assess how a changed flow would affect fish ecology. Klumpströmmen is unusual compared to many other rivers and river sections in Sweden in that it has not been cleaned for timber floating. Therefore the riverbed is more heterogenous than the average river, even when compared to unregulated rivers. This aspect has a positive influence on potential biological productivity. In addition, Klumpströmmen is an outlet stream and this contributes to productivity, both concerning fish- and benthic fauna. However, the outlet effect is not unique to Klumströmmen. In fact, many bypass channels are close to dams, which mean that outlet effects should be common in these types of systems.

Our scenarios are traditional constant flow regimes, with a low winter flow and a higher summer flow. Alternatively, we could have suggested a variable flow to mimic a natural flow regime.[5] But the power station in Dönje has a maximum capacity that is exceeded at high flows. Therefore, the flow in Klumpströmmen is expected to become variable even in the absence of a legislated two-level flow. The most important aspect that makes our scenario "tick"is that Klumströmmen is a natural side channel or bypass channel. Before the hydroplant was constructed, Klumpströmmen had about 25% of the total flow; during extreme highflows excess water can be spilled into the main channel to protect the smaller Klumpströmmen from severe disturbance during extreme high flow events.

Six flow regimes ranging from the present minimum legislated winter flow of $0.25\,\mathrm{m^3s^{-1}}$ to pristine low summer flows of $41\,\mathrm{m^3s^{-1}}$ were evaluated at the regulated, but riverbedwise uniquely pristine stretch Klumpströmmen. These scenarios were generated "live", by asking the power company to release certain amounts of water during prescribed periods of time. Thus, in June 2008 when the minimum required flow is 10, 21 and $41\,\mathrm{m^3s^{-1}}$ were released from the plant. In May 2007, during the minimum legislated winter flow, the team looked at flows lower than $10\,\mathrm{m^3s^{-1}}$, but higher than $0.25\,\mathrm{m^3s^{-1}}$.

Compared to the minimum winter flow, with no suitable habitats for grayling and brown trout, the area of suitable spawning and juvenile habitats at $3\,\mathrm{m^3s^{-1}}$ were estimated to 3 ha, and 0.5 ha for adults. At the present legislated minimum summer flow of $10\,\mathrm{m^3s^{-1}}$ the suitable habitats covered an area of about 3 ha for juveniles and 4 ha for adults. At higher flows these areas increased proportionally less than

---

[5]We did experiment with such a scenario in the focus groups, but found that it is not easy to explain such a scenario to a respondent.

the corresponding increase in the flow ($21\,m^3s^{-1}$; juveniles 4 ha, adults 5–6 ha and $41\,m^3s^{-1}$; juveniles 6 ha, adults 8–9 ha).

In relation to reference data, snorkeling in the river section indicated low fish densities, averaging 0.39 grayling per $100\,m^2$, while no brown trout was observed. By electro-fishing, an average of 0.14 brown trout per $100\,m^2$ was estimated, which in relation to reference data indicated a very low density. Based on the collected information, the current flow regulations in the section seem to impair both the grayling and brown populations. With environmentally adapted flows implying c. 1.5 month earlier start and ending of minimum summer flow and increasing the minimum winter flow from 0.25 to $3\,m^3s^{-1}$, the salmonid density was predicted to increase by a factor of 3–6 compared to that estimated in the field, see [154] and [155]. These predictions, combined with other information about the possible impact of the scenarios, were included in the Contingent Valuation study, to which we now turn.

5.4 Contingent Valuation Study of Improved Downstream River Basin

The Contingent Valuation study targeted respondents in the Bollnäs municipality, but the survey also gave some information about WTP^{RB} in neighboring municipalities. It was preceded by the focus group analysis described below and the on-site ecological assessment described above. These two studies were the basic ingredients in the scenario development. An important step in the development of a survey is to conduct focus groups to test how a general audience interprets the survey, subsequent steps include survey revisions, a pilot study, and a final study. This last step was omitted here, because the web-panel was essentially exhaustive in the Bollnäs area and we did not have resources to undertake additional interviews. Specifically, our focus groups were designed to ensure, among other things, that the proposed environmental changes and the hypothetical valuation scenario described in the survey were understood by respondents.

5.4.1 The Focus Group Study

In July 2008 the economics team in the project held six focus groups in the city of Bollnäs, meeting with a total of 39 individuals. The focus groups of three to seven people were moderated by two researchers from the Swedish University of Agricultural Sciences (SLU) and lasted approximately 75 min. Respondents received a cash payment of SEK 250 to compensate for their time. Respondents were asked to answer all survey questions and to write notes, questions or comments in the margin to indicate confusion or misunderstanding. At designed places in the survey, respondents were asked to stop and discuss their reaction to the questions. The moderators asked probing questions to determine whether respondents understood the key aspects of the valuation scenario. The ensuing discussion resulted in valuable insights into how the survey was interpreted and provided useful information for future revisions.

Below we highlight some of the key points that arose from the 4 days of focus group testing.

- Our valuation scenario that the municipality of Bollnäs buys back the water rights (fallrättigheter) from Fortum (the owner of the Dönje power plant) was generally well understood by respondents and seems to represent a reasonable valuation scenario. While there were protests against Fortum as a company (see bullet below), the protest was not against our idea of buying back the water right itself, which most respondents agreed was a reasonable way to phrase the question.
- Some respondents refused to give a value for improved water flows, which seemed to indicate a protest against Fortum, rather than a zero value for the resource. These respondents insisted that the company responsible for causing the problem should pay to fix it, not the general citizens. This type of protest (or personal bias) against the company that contributes to the perceived environmental problem is a tricky issue in these types of valuation surveys. We addressed this through improved survey design and explanation of the survey's purpose to respondents.
- The pictures shown in the survey were important to most respondents, but they required some improvement to ensure consistency across comparison, that is, all aspects of the photographs need to be identical, except the water level. One respondent said the picture of a scenario involving $30\,\mathrm{m}^3\mathrm{s}^{-1}$ looked like a natural river as she remembers the unregulated river that she grew up next to as a child. But others did not see a big difference in the pictures of different water levels. One person said that the $30\,\mathrm{m}^3\mathrm{s}^{-1}$ looked like a dangerous flood level to her, which may be due to the fact that she has never seen how the river looked prior to the dam ($30\,\mathrm{m}^3\mathrm{s}^{-1}$ is well below the lowest average monthly flows pre-dam) or may be because of the plants that have grown into the riverbed and make it look like a flooded river at flows higher than $10\,\mathrm{m}^3\mathrm{s}^{-1}$. This type of reaction to the pictures can be addressed through photo-manipulation that removes the plants and shows how the river will look after several successive years of higher water levels. These comments were valuable because they show that we needed to improve our description of this river under higher future flows. Most people who live here have never seen higher flows in a natural setting and therefore do not understand what they are "purchasing" in the survey.
- Some respondents were not clear whether our scenario was hypothetical or whether their vote for improved flows might impact the provision of other services offered by the municipality (e.g., school, hospital). We explain carefully in the survey that our proposal would require a real (not hypothetical) payment from citizens in Bollnäs in order to provide the additional funds for the purpose of higher water flows.

- Some respondents worried that their electricity bills would be higher in addition to the cost of the proposal in the survey or that the proposal would lead to electricity production from other "dirty" sources, such as nuclear. We explain in the survey that the change at Dönje power station is a marginal change that is unlikely to lead to noticeable impacts on prices.
- Many respondents indicated that the three proposals for improved water flows failed to explain all of the potential environmental and non-environmental impacts. For example, respondents said they had trouble voting for or against our proposals because they wanted to know more about how the hypothetical future change in water level would affect the ecosystem, their recreational activity (paddling, swimming), future fish tourism activity around Bollnäs, water levels in the up- and downstream lakes, the type of fish species likely to be found in Bollnäströmmarna, or the fishing access and fishing pressure in these areas. We made significant improvements in this area in the survey.
- Some respondents indicated that the survey focused too much on the benefits to the number of fish in the river, which was less important to non-fishermen. In the survey we use pictures to describe how improved water flows lead to improved water and habitat quality and, hence, a healthier ecosystem in Bollnäsströmmarna.
- Several respondents found it difficult to estimate their own value of increased water flow and, instead, wanted to know what it would cost Fortum in terms of foregone electricity. If they knew this figure, then they thought it would be easier for them to determine whether they should vote for the proposal. We revised our explanation of the purpose of the survey to make clear that we want to estimate a value for improved flows and that this value will subsequently be weighed against the costs to society in terms of foregone hydropower profits.

These were the main insights provided by the focus group study. The focus group study as well as other local contacts in the area provided additional knowledge and information that was used in designing the Contingent Valuation experiment and the other items constituting the empirical assessment of the two considered projects/scenarios.

5.4.2 The Pilot Study

Given the insights obtained from the focus-groups, we then proceeded to revise our survey instrument. A web-based sample was obtained providing 136 completed surveys for the Bollnäs municipality, and 200 completed surveys in total. The $200 - 136 = 64$ questionnaires not from the Bollnäs municipality came from households in neighboring municipalities (five cases had invalid zip codes and were deleted from the analysis of residence), see Table 5.1.

Thus, a majority of the respondents reside in Bollnäs municipality and those who do not live in neighboring municipalities. While there is some interest on how

Table 5.1 Area of residence and number of respondents in Bollnäs municipality and neighboring municipalities

Respondent's residence	# of respondents	Municipality of residence
Alfta	4	Ovanåker
Arbrå	13	Bollnäs
Bergvik	4	Söderhamn
Bollnäs	87	Bollnäs
Edsbyn	6	Ovanåker
Enviken	4	Falun
Holmsveden	1	Söderhamn
Järvsö	11	Ljusdal
Kilafors	12	Bollnäs
Ljusne	4	Söderhamn
Marmaverken	1	Söderhamn
Rengsjö	11	Bollnäs
Sandarne	1	Söderhamn
Segersta	1	Bollnäs
Stråtjära	1	Söderhamn
Söderhamn	18	Söderhamn
Trönödal	1	Söderhamn
Vallsta	12	Bollnäs
Vallvik	2	Söderhamn
Viksjöfors	1	Ovanåker

residents outside of Bollnäs did vote, the hypothetical referendum is based on the assumption that the respondent belongs to Bollnäs municipality.

We then undertook a straightforward representativity analysis of the sample, which is available on the book's web material site. This analysis shows that there are only small differences between our sample and the population of Bollnäs. An interesting fact is that we had a slight underrepresentation of young households and a slight overrepresentation of old, which might not be expected in a web-survey.

Let us turn to the interval WTP-questions, which was our chosen elicitation mechanism.

5.4.3 Interval WTP Questions

A key issue in any survey is the elicitation architecture, the way of eliciting information from the respondent. A number of disadvantages have been demonstrated with the standard elicitation methods in surveys.[6] For example, open-ended questions tend to depress response rates, while the close-ended question necessarily gives

---

[6]A compact survey of many issues in survey research, in particular regarding response errors and biases across formats is given in [136].

much more limited information. Each of these question types can be further subdivided into several categories. For example, closed-ended questions can be based on a Likert-scale (e.g. from "strongly disagree" to "strongly agree"), a multiple choice (e.g. 'circle one of the following alternatives'), an ordinal question (e.g. 'rank the following items from 1 to 5'), and a binary question ("are you willing to pay X USD for this public good" (yes, no)). There are also several variants of the open-ended questions, such as "How much are you willing to pay for this public good?" or "How did you make that choice?". Obviously, the choice between the open and closed-ended questions is not straightforward, because they have advantages and disadvantages in different situations, see, for example, [55].

In the non-market valuation literature, the discussion about the "best" way to elicit individual's willingness-to-pay is still ongoing. In the early 1960s, bidding game and open-ended questions were mostly used. Practitioners then turned to payment cards (a version of bracketing) and binary valuation questions, in both cases suggesting ranges of prices or one price for a proposed environmental change. The binary question has become the gold-standard, especially variations of this idea, including double bounded (two binary questions) and adding a zero valuation opt-out to this "yes-no" type question.[7] When an individual is offered a price to reject or accept, there is a tendency for anchoring around the price; while there is no intended information content the individual may well anchor his valuation around the suggested price. In the psychological literature, this phenomena was documented in the beginning of the 1970s.[8] Anchoring also seem to affect the popular bracketing approach, in which the respondent is to select one out of a number of given intervals, see [187].[9]

Recently, there has been a surge of interest in valuation uncertainty. The currently most popular approach in Contingent Valuation is a payment card containing several different costs for a public good, combined with a question about how certain the respondent feels about paying a certain cost (e.g. "definite yes", "probably yes", "probably no" and "definite no"). Recent analysis shows that including such uncertainty-assessments in the survey instrument may affect the estimate of valuations. See the review of this literature and the analysis in [25].

There is a growing interest in alternative elicitation methods based on intervals. Consider, for example, The University of Michigan Health and Retirement Study (HRS) 2008 survey which surveys more than 22,000 Americans over the age of 50

---

[7]The NOOA-panel on contingent valuation has been influential in this regard. Focussing damage assessments of oil-spills in a litigation context, the panel concluded that contingent valuation serves as a useful starting point if, inter alia, binary valuation questions are used.

[8]Reference [134] and [172]. A candid review of these early and very influential papers is provided by [122]. In particular, Kyle's review suggests that the extent to which individual's use heuristics for substantive decisions is unclear.

[9]For a general account of how "questions shapes the answers", see [160].

every 2 years since 1992. In probing the individual's answers to one question about expectations, the survey uses the following set-up[10]

> When people are asked to give a numerical response, like percent chance, sometimes they give exact answers and sometimes they give rounded or approximate numbers ....
> When you said ... percent just now, did you mean this as an exact number or were you rounding or approximating?
> What range of numbers did you have in mind when you said ... percent? (enter a range) between (min percent) AND (max percent).

As noted in [133] a basic problem is that when a respondent is to report a point, he sometimes rounds it to describe a sentiment that really is an interval. Reference [133] present an approach to deal with such intervals, which is different from the one suggested in [11] and described in more detail below. The difference arises partly because they ask the respondent to state a point or an interval, not both.

Reference [11] argue that self-selected intervals provide a richer picture of response uncertainty, potentially increase response-rates and maintain a link to recent ideas on coherent arbitrariness (see [76]). We implemented this approach in the present study by allowing the respondents to choose between a point and an interval. Thus, if they could not answer the question with a point estimate, as basic economic theory does suggest, they were then asked to submit the answer as an interval. In other words, we assume that true WTP is either stated as a point or contained in the reported interval.

The self-selected intervals are connected to two strands of literature, one in the survey research literature, the other in the statistics literature. From the survey research angle, it is convenient to think about this idea in terms of approaches to elicit a probability distribution $F[x] = P[X \leq x]$ (for a survey, see e.g. [63]). In practice, respondents can ordinarily not state $F[x]$ with any precision, and practice involves asking for a few points of the distribution or various summary statistics. Reference [63] distinguishes between the *fixed interval method* and the *variable interval method*. In the first case, the respondent is asked to assess the probability that X is within a set of intervals proposed by the investigator and the constraint that probabilities sum to one is imposed. In the second, the respondent is asked to state the upper and lower quartiles for X, the maximum amount of tomorrow's precipitation, for example. The interval has a specified probability, e.g. 50% chance that the interval will cover the true value. These ideas are related to the self-selected intervals we use here.

In the statistical literature, the concept of self-selected interval is closely related to interval-censored failure time data in survival analysis. Censoring mechanisms can be quite complicated and thus necessitate special methods of treatment. Different types of interval-censored data have been studied. Reference [89] and [190] considered the partly interval-censored failure time data where observed data include both exact and interval-censored observations on the survival time

---

[10]see http://hrsonline.isr.umich.edu/modules/meta/2008/core/qnaire/online/16hr08P.pdf, HRS 2008, SECTION P, EXPECTATIONS, PAGE 11, FINAL VERSION, 9/28/2009.

of interest.[11] Reference [96] introduced the concept of "middle censoring" which occurs when an observation becomes unobservable if it falls inside a random interval. However, standard methods for analyzing interval-censored and middle-censored data assume, implicitly or explicitly, that the censoring intervals are independent of the exact values. This is an assumption that may be questioned.

Researchers in the ELFORSK-project on which the bulk of this book is based, have developed and investigated some new methods to estimate mean WTP for the situation when self-selected intervals are allowed. In reference [11] it is argued that there are two different probabilities involved. One probability is related to the conditional probability of finding WTP in a given interval; the second probability is related to the probability of choosing a particular interval. Here it is assumed that the censoring intervals follow a bivariate distribution with finitely many support points. Assuming that the respondents tend to state rounded intervals from a finite set, this is a reasonable assumption. In reference [11] it is furthermore assumed that all respondents give self-selected intervals and consistency is proved. We will explain this estimator in more detail below, but let us look at some alternatives first.

A number of suggestions are based on maximum likelihood principles. The most popular non-parametric maximum likelihood estimators are based on [171]. A Turnbull type likelihood is based on the assumption that the censoring mechanism is random from the individual's point of view. Thus, each interval is viewed as if the individual choose that particular interval among other presented intervals. This is not the case here, because the individual can self-select any interval of choice. Reference [76] propose a linear approximation of the distribution function over an interval. For each interval, the probability that WTP is greater than the stated lower bound is 1, while the probability that it is greater than the upper bound is zero. Reference [76] suggest a linear approximation such that $E(WTP) = \frac{2}{3} \times L + \frac{1}{3} \times U$, where L is the lower bound and U is the upper bound. This is a slightly more conservative estimator, compared to using the mid-points (i.e. using the weight $\frac{1}{2}$ for each bound).

Reference [48] assumes that some respondents give exact answers, some intervals and a third group answer both an exact value and an interval. It is assumed that the censoring intervals follow a bivariate distribution with finitely many support points. For this situation he derives a non-parametric maximum likelihood estimator. In a simulation study he compare that estimator with the [96] estimator and finds that the latter has a tendency to give biased estimates if the respondents has a tendency to let the exact value be close to one of the endpoints. The proposed estimators by [48] do not seem to share this problem.

In reference [188] three different approaches to estimate the mean willingness to pay are studied. It is assumed that some respondents give an exact value and that some give a self-selected interval. First they consider the nonparametric (the Jammalamadaka and Mangalam estimator) and a parametric approach (assuming a Weibull distribution) where the intervals are treated as if the respondent gives an exact value but we cannot observe it. Reference [188] give a different interpretation

---

[11]There is also a literature using Bayesian analysis. For an application in our context, see e.g. [54]

of the intervals. They argue that the respondents answer provide information about his/her uncertainty about what would be a reasonable value of WTP. They assume that, because of a number of uncertainties, instead of giving an exact value the answer is given by a random variable having a certain distribution. Furthermore, they assume that a respondent cannot give the answer as a distribution, but approximate it by giving the lower and upper value of the interval. In essence, they consider two types of individuals. Type 1 reports a point value, say X_i, while type two has $X_i + \epsilon_i$, where ϵ_i follows a triangular distribution. The variance of ϵ_i is proportional to the square of the interval length. Thus, the idea is that respondents provide the lower and upper bounds of their own distribution and the length of the reported interval signals the uncertainty. They also find that Jammalamadaka and Mangalam estimator underestimates WTP in our data, and propose the triangular distribution with mode at the right endpoint. Let us now return to the Belyaev-Kriström proposal.

5.4.4 Belyaev-Kriström Estimator for Self-Selected Intervals

The estimator is based on the following three assumptions:

Assumption 1 *Each individual has one true point of compensating variation, but might not be aware of the exact location of this point.*

Assumption 2 *The true points of compensating variation are independent of question mode (open-ended, a self-selected interval or a choice between the two) and the question mode does not change the true points.*

Assumption 3 *The true points, the stated points and the intervals of compensating variation corresponding to different respondents in a sample are values of independent identically distributed (i.i.d.) random variables.*

Assumption 1 is the basis for the approach. Observe that in standard utility theory, the individual does know his valuation as a point. We say in the scenario that "The change will improve fishing conditions, water ecology and landscape aesthetics" and give some details on estimated fish population changes. We then display in photographs how the environment will change, but the scenario must by necessity involve some uncertainties. In principle, the self-selected intervals allow the individual to map out this implicit uncertainty. So if the fish population actually does increase by 75%, then he will pay some amount, but if it increases by 125% he might pay more. Assumption 2 is, perhaps, the strongest. It should not matter to the individual's valuation if we ask him an open-ended question or a self-selected interval question. Assumption 3 seems innocent, but if we are applying contingent valuation in a very small neighborhood, the assumption might not hold.

We now sketch the statistical model and refer the reader to [11] for a detailed presentation and proofs. Let the collected statistical data consist of n stated rounded intervals $\mathbf{y}_i = (y_{Li}, y_{Ri}]$ containing unobserved true WTP-points of compensating variation $x_i \in \mathbf{y}_i$, $i = 1, \ldots, n$. Let $\mathscr{U}_m = \{\mathbf{u}_1, \ldots, \mathbf{u}_h, \ldots, \mathbf{u}_m\}$ be the list with all different intervals $\mathbf{u}_h = (u_{Lh}, u_{Rh}]$, $u_{Lh} < u_{Rh}$, $\mathbf{u}_{h_1} \neq \mathbf{u}_{h_2}$, $h_1 \neq h_2$, and for each $i \in \{1, \ldots, n\}$ there is at least one $\mathbf{y}_i = \mathbf{u}_{h_i}$, i.e. the stated WTP-intervals $\{\mathbf{y}_1, \ldots, \mathbf{y}_n\} = \{\mathbf{u}_{h_1}, \ldots, \mathbf{u}_{h_n}\}$. Let t_h be the number of cases when $\mathbf{y}_i = \mathbf{u}_h$. and we can then write the data as a list $\mathbf{dat}_m = \{\{\mathbf{u}_1, t_1\}, \ldots, \{\mathbf{u}_h, t_h\}, \ldots, \{\mathbf{u}_m, t_m\}\}$, $\sum_{h=1}^{m} t_h = n$.

By Assumption 3 we consider $\{x_1, \mathbf{y}_1\}, \ldots, \{x_n, \mathbf{y}_n\}$ as values of i.i.d. pairs of random variables (r.v.s) $\{X_i, \mathbf{Y}_i\}$, $\mathbf{Y}_i = (Y_{Li}, Y_{Ri}]$, and let $\{\mathbf{Y}_i \leq \mathbf{y}\} = \{Y_{Li} \leq y_L, Y_{Ri} \leq y_R\}$. Their d.f.s $F[x] = P[X_i \leq x]$ and $G[\mathbf{y} \mid x] = P[\mathbf{Y}_i \leq \mathbf{y} \mid X_i = x_i]$ are unknown. By Assumption 1 we have $X_i \in \mathbf{Y}_i$.

The stated rounded intervals $\mathbf{u}_h = (u_{Lh}, u_{Rh}]$, $h = 1, \ldots, n$, can overlap and their union is contained in the support of the distribution of the r.v.s X_i, $i = 1, \ldots, n$. The ith respondent states that the true point of compensating variation x_i belongs to an interval $\mathbf{u}_h$, e.g. "$x_i \in \mathbf{u}_h''$, $u_{Lh} < x_i \leq u_{Rh}$.

Let $\mathscr{V}_k = \{\mathbf{v}_1, \ldots, \mathbf{v}_k\}$ be the division generated by the set of intervals $\mathscr{U}_m$, i.e. $\mathscr{V}_k$ is the collection of disjoint intervals $\mathbf{v}_j = (v_{Lj}, v_{Rj}]$ and each $\mathbf{u}_h = \cup_{j \in \mathscr{C}_h} \mathbf{v}_j$, where $\mathscr{C}_h = \{j : \mathbf{v}_j \subseteq \mathbf{u}_h\}$ is the set of all indices of division intervals which are subsets of $\mathbf{u}_h$, $h = 1, \ldots, m$. For each $j = 1, \ldots, k$ we define the set $\mathscr{D}_j = \{h : \mathbf{v}_j \subseteq \mathbf{u}_h\}$, i.e. h belongs to $\mathscr{D}_j$ if and only if $\mathbf{v}_j \subseteq \mathbf{u}_h$. By d_j we denote the number of $h \in \mathscr{D}_j$, i.e. d_j is the size of $\mathscr{D}_j$. The division $\mathscr{V}_k = \{\mathbf{v}_1, \ldots, \mathbf{v}_k\}$ may be considered as a kind of bracketing generated by respondents due to roundings.

To illustrate these concepts, consider a small subset of the data listed in the appendix, see Table 5.2.

Thus, the division intervals are $(50, 100]$, $(100, 200]$ and t_h is 3 and 6 respectively. For example, for an individual with self-selected interval $(50, 200]$ we splice the interval into two parts. With more data, each self-selected interval tends to have more divisions.

The individual is choosing an interval from a finite set and we now define related probabilities. On any division interval $\mathbf{v}_j$ the probability to state $\mathbf{u}_h$, $h \in \mathscr{D}_j$, is

$$w_{hj} = P[``_i \in \mathbf{u}_h'' \mid X_i = x_i \in \mathbf{v}_j], \quad \sum_{h \in \mathscr{D}_j} w_{hj} = 1, \ w_{hj} > 0. \tag{5.5}$$

Table 5.2 Subset of interval data from Appendix F for illustration of interval data

Respondent ID	Lower	Upper
4	100	200
5	100	200
11	50	200
12	50	200
16	50	200
17	100	200
30	100	200
38	100	200
45	100	200

Using the law of total probability, the ith respondent states an interval $\mathbf{u}_h$ with probability

$$w_h = P[``x_i \in \mathbf{u}''_h] = \sum_{j \in \mathscr{C}_h} w_{hj} F[\mathbf{v}_j], \quad h = 1, \dots, m. \qquad (5.6)$$

We have little or no information about how respondents choose their intervals, but several plausible assumptions exist. Reference [12] propose five different behavioral assumptions, all of them portraying how the individual chooses the interval. For example, if the WTP-point happens to be in a particular division, the individual might choose to report the interval in which this particular division is the last.

We will assume a parametric distribution function.[12] Let us use the Weibull d.f. $W(a, b) = F_{ab}[x]$, which we write as $1 - e^{-(x/a)^b}$, where a is the scale parameter, and b is the shape parameter. We thus approximate the d.f. of unobserved true WTP-points values $x_1, \dots, x_m$ by a distribution from the Weibull family $\mathscr{F}_W$ of distributions. The properties of the estimator are given in the following theorem.

Theorem 1 *If Assumptions 1–3 hold and the distribution of true WTP-points is $W(a, b)$ then a consistent ML-estimator $\hat{\theta}_n = \{\hat{a}_n, \hat{b}_n\}$ exists and its accuracy can be consistently estimated by resampling as $n \to \infty$.*

Proof. See [11].

The log-likelihood in the case of our Weibull model is,

$$\text{llik } W[a, b \mid \mathbf{dat}_m] = \sum_{h=1}^{m} \frac{t_h}{n} \text{Log} \left[\sum_{j \in \mathscr{C}_h} \hat{w}_{hj} \left(e^{-(v_{Lh}/a)^b} - e^{-(v_{Rh}/a)^b} \right) \right]. \qquad (5.7)$$

In order to obtain a bit more intuition about this estimator, it is instructive to compare it with the widely known Turnbull approach. A parametric version of the Turnbull estimator entails maximizing

$$\text{llik } WB[a, b \mid \mathbf{dat}_m] = \sum_{h=1}^{m} \frac{t_h}{n} \text{Log} \left[\left(e^{-(u_{Lh}/a)^b} - e^{-(u_{Rh}/a)^b} \right) \right]. \qquad (5.8)$$

This is simply the product of the probability of finding a value in each given interval, and this is a basic construction in survival analysis when data are interval-censored. The difference between the self-selected statistical model and the Turnbull approach can be seen from Eqs. 5.7 and 5.8. The main difference is that Eq. 5.7 includes a sum over the divisions, while Eq. 5.8 has a much simpler probability statement. The expressions encapsulate the key difference between the way the data are generated. We interpret Eq. 5.7 to mean the individual can freely choose an interval in a finite set, while Eq. 5.8 portrays the likelihood when the individual is presented with

---

[12]For a discussion of a non-parametric ML see [11].

certain brackets by an investigator. The "cost" of this freedom from a computational point of view is displayed in Eq. 5.7.

5.4.5   Valuation Question

As noted above, the WTP-question was formulated as a kind of referendum in the Bollnäs municipality. This follows a long-standing tradition of local referenda in some countries/states (e.g. Switzerland/California) in general and on water issues in particular. For example, the *Wyoming Preserve Minimum Instream Water Flows Initiative*, May 23, 1982 in Wyoming, USA mandated leaving a certain level of water in streams so that natural fisheries could thrive. The initiative was set for the 1986 ballot, but did not make it.[13]

In the present study the question was formulated in the following way:

> **WTP-question introduction** *It has become more common in Sweden that those who are affected by local environmental issues are able to vote in local referendums. The following proposal can be viewed as such a local referendum. It is thought to be held among inhabitants of Bollnäs municipality. The purpose of this question is to shed some light on how the average citizen of Bollnäs values a potential change of the water flow in the Bollnäs streams. The change will improve fishing conditions, water ecology and landscape aesthetics. At present, the water rights are owned by the Fortum company. This means that Fortum has the right to produce electricity at the Dönje plant. Suppose that the only possible way to increase the water flow in the Bollnäs streams is to buy back those water rights, by means of a joint action among Bollnäs citizens.*

We then described the winter-season proposal.

> **Proposal to change of the winter season water flow in the Bollnäs streams.** *The* **Proposal** *entails an increase of the winter season water flow from 0.25 to 3 $m^3 s^{-1}$. There will be no change in the summer season water flow. This means that the water flow will increase from the power station to Varpen (note: this was described in a map not included here) The proposal is depicted in a series of pictures in the sequel. The total costs for the* **Proposal** *is not known with certainty at the present time. Suppose that the referendum is held when the cost has become known. If a majority supports the proposal, it will be undertaken. A "yes" entails each household paying a given sum over a period of 5 years.*

---

[13]It was argued that the Wyoming State Legislature in 1985 accomplished substantially the same objectives. Source http://ballotpedia.org/wiki/index.php/Wyoming_Preserve_Minimum_Instream_Flows_Initiative_(1986)

A series of pictures were introduced in order to depict SCENARIO 1, see Figs. 5.1 and 5.2. The drawing was necessary, because we have no photographs of the change during the ice-period. For SCENARIO 2, a picture was added showing the summer season change, see Fig. 5.3. Indeed, the difference between the status quo and the scenario, from 10 to $20\,m^3s^{-1}$ in the summer season, is quite small. Consequently, we expect rather similar valuations of the two scenarios, an expectation that was fulfilled, as is shown below.

Fig. 5.1 The status quo and the change in the ice-free winter period (SCENARIO 1)

Fig. 5.2 The new winter season situation under the ice-period (Note: "Dagens läge på vintern" = Todays situation in the winter. "Situationen på vintern vid ökat vattenflöde" = Situation in the winter at increased water flow)

Fig. 5.3 The status quo and the change in the summer period (SCENARIO 2)

The next question included an opt-out possibility, i.e. a zero WTP possibility. Those who rejected this option could then state their WTP. If an individual was not able to state WTP as a point he was asked to state it as a freely chosen interval.

5.4.6 Estimating WTP

First, it is of interest to consider the "size of the market". Because the scenario entails payment responsibility only for those living in the Bollnäs municipality, we focus on these respondents in the sequel. Around 62% report $WTP > 0$ and 38% report $WTP = 0$. Thus there is a majority in favor of the project.

We begin the analysis by summarizing the data used in Table 5.3.

The table makes it plain that we do have a sparsity of data, and, again, our analysis here is only to illustrate how to undertake a modern cost-benefit analysis of water use conflicts. In order to obtain average WTP, we simply add the proportion of zero WTP in the data, those respondents who claim not to be "in-the-market". In so doing, we need to handle the six respondents who reported an interval (0,upper). We interpret them to be in the market.[14]

The figure that we use in the empirical CBA is taken from Table 5.3 by weighting in the zero answers. It is calculated as $\frac{36}{135} \times 482.25 + \frac{47}{135} \times 495.08 \approx 300$, where 495.08 is the average of the mid-points. In Table 5.4 we collect estimates from using the [96] (JM03) estimator, a Weibull distribution, the midpoints and [11].

The [96] estimator is a non-parametric estimator of the underlying distribution of WTP and is, as noted, developed for data that are middle-censored. It has been shown that in [96], the population distribution F of WTP is given by

$$F(t) = P(t) + \int \frac{F(t \wedge U) - F(t \wedge L)}{F(U) - F(L)} dQ(L, U), \qquad (5.9)$$

Table 5.3 Descriptives of WTP, conditional on $WTP > 0$

WTP	n	Unique	Mean	0.50	0.75	0.90	0.95
Lower	47	13	214.3	100	100	540	1,000
Point	36	13	482.2	250	500	1000	1,375
Upper	47	19	775.9	500.0	1,100.0	1,700.0	2,350

---

[14] An extreme outlier has been deleted from the data. The stated WTP exceeded the individual's reported income by a significant amount. While it is possible, in principle, for this individual to borrow on the market to finance his contribution, we assume that the reported sentiment does not reflect true WTP. The reported value is 1779 standard deviations from the mean. If the WTP-distribution is normal with the estimated means and standard deviations, the probability of finding such an observation is much less that 10^{-22}. We also deleted an observation where an individual said he (or she) was willing to pay, but refused to state a point or an interval.

where $P(t)$ is the distribution of the point data and Q the distribution of the intervals of the censored observations, U and L are the upper and lower endpoints of a reported interval. Intuitively, if there are no intervals, then $F(t) = P(t)$. If there are intervals, then one must consider the conditional probability for finding the censored observation covered by an interval of size (L, U). The non-parametric estimate of F can be obtained by the EM-algorithm, as shown in Appendix F.

The parametric Weibull estimator for the data that includes the WTP-points, given positive valuations, is just the likelihood Eq. 5.8 with the density added. Using maximum-likelihood, we find that $a \approx 0.86$ and $b \approx 389.72$ for the best fitting behavioral model (see Appendix F.).

The code that we used is listed in Appendix F. Because the mean of a Weibull is $b\Gamma(1 + \frac{1}{a})$, these parameter estimates are sufficient to calculate the mean.

For the Belyaev-Kriström estimator an R-package called `iwtp` has been developed. It is available at CRAN. In the appendix, we explain how to use it to replicate the results reported here. For the chosen Weibull model (behavioral model 5, see Appendix Sect. F.4.2 for exact definitions), we obtain the following estimates of the distribution function in Fig. 5.4.

The non-parametric estimator is just the empirical distribution function for the upper bounds. Mean WTP from the Weibull model is 449.41, with the scale parameter estimated at 410.06 and the shape parameter at 0.84 (i.e. $410.06 * \Gamma(1 + 1/0.84)$). We used re-sampling (implemented in the `iwtp` program) and obtained the following quantiles of mean WTP; $5\% = 306.12$, $50\% = 438.14$, $95\% = 562.30$. The `iwtp` program has additional capabilities, such as the ability to estimate a mixed Weibull/exponential for self-selected data and assessing the models by specialized quantile-quantile plots and more. However, we will leave those extensions to a case when we have more data.[15]

The mean of the points is 482.25. So, if we want to get an overall mean we also add the zeroes to get $(47/135) * 449.41 + (36/135) * 482.25 = 285$. This is fairly close to the number 300 that we use in the CBA.

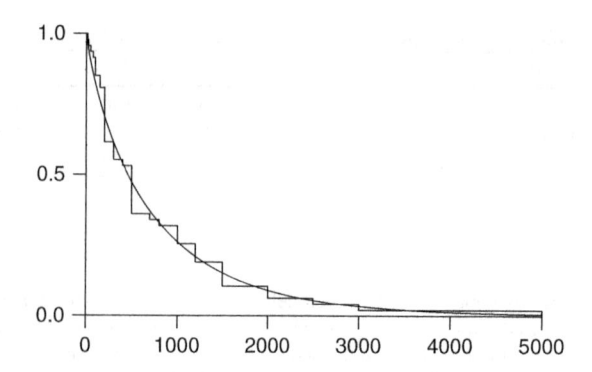

Fig. 5.4 Non-parametric and parametric estimator of the distribution for WTP according to the Belyaev-Kriström estimator

---

[15]The mixed Weibull-exponential model gave rather similar results, with an estimated mean of 444.

Table 5.4 Estimated mean WTP for different estimators

Estimator	JM03	Weibull	Midpoints	BK09
Mean	286	258	301	285
Standard error	45	33	41	See text

As we can see from Table 5.4 the estimates are rather similar, although we must emphasize that the number of observations is very small. To repeat, for ease of replication we use the rounded number 300 as our benefit estimate in the empirical CBA. Finally, we note that [188] report very similar estimates using different estimators but the same data set with their means ranging from 258 to 334 depending on the specific assumptions made.

5.5   The Benefits of Smoother Downstream Flow

As noted, it is possible, although by no means certain, that the proposed scenarios will change river flow volatility. The plant will operate within a new set of constraints with its regulation services reduced. In turn, this could affect flow volatility, depending on how Fortum replaces the reduced regulation service at the Dönje plant. If Fortum does re-optimize, we can have effects both up and downstream the plant. The company may also use another river or a different generation technology for replacing the lost regulatory service. In our particular application we expect the following, given discussions with traders at the Fortum company. SCENARIO 1 entails no change in the regulatory regime, and therefore it should not change downstream flows relative to the current situation. The main reason is that the change is too small to warrant significant changes of the planning. There is a downstream tributary (Galvån) that fluctuates between 2 and $8\,\mathrm{m^3s^{-1}}$, that affects the downstream situation in the same way as an increased bypass flow at the plant. SCENARIO 2 is more complex, because it involves a more substantial change in water spill, thus potentially creating re-bound effects on upstream and downstream power stations. Because the price is higher during the day, the owner of the Dönje plant strives to save water during the night to sell during the higher priced day period. Squeezing an extra $10\,\mathrm{m^3s^{-1}}$ from Dönje, while simultaneously building up water storage at the Dönje dam entails releasing water from an upstream plant during the night. This intertemporal re-allocation means producing more at lower prices and thus changing the upstream flow during the night. In short, the intertemporal re-allocation maps into shortening the daytime production at Dönje. But overall, the best approximation seems to be that the downstream effects are small. Therefore, in the empirical CBA the term WTP^B reflecting the benefits of smoother downstream flow is set equal to zero.

5.6 Externalities from Replacement Power

Because our project potentially entails a small reconfiguration of the way electricity is generated, we briefly discuss approaches to quantify the new vector of unpriced externalities, following any replacement of the lost power. There is some controversy regarding the concept of marginal electricity, but here we stick to the standard interpretation. First of all, we note that if externalities are correctly taxed, or traded in an optimally designed emission trading system, we can disregard them in a CBA of a small project. The argument is exactly the same as with marketed goods; a correct price measures the social value of the good. Externalities from power generation have been subject to a massive number of studies, but to fix ideas in Fig. 5.5 we present one out of many summaries of the extant literature (for additional information, see e.g. the extensive EXTERNE project).

As can be seen, the variation is substantial, from virtually zero to roughly EUR 80 per MWh, depending on the source. While these estimates provide interesting information, we only consider externalities that are not priced. The Swedish energy system is virtually fossil-free and the most interesting aspect is replacement by for example Danish coal-power since the marginal producer on the Nord Pool spot market typically is a coal-fired plant; see Fig. 4.1 in Chap. 4. Suppose the scenarios maps into replacing the lost (3.7, 14.3) GWh with Danish coal powered electricity.

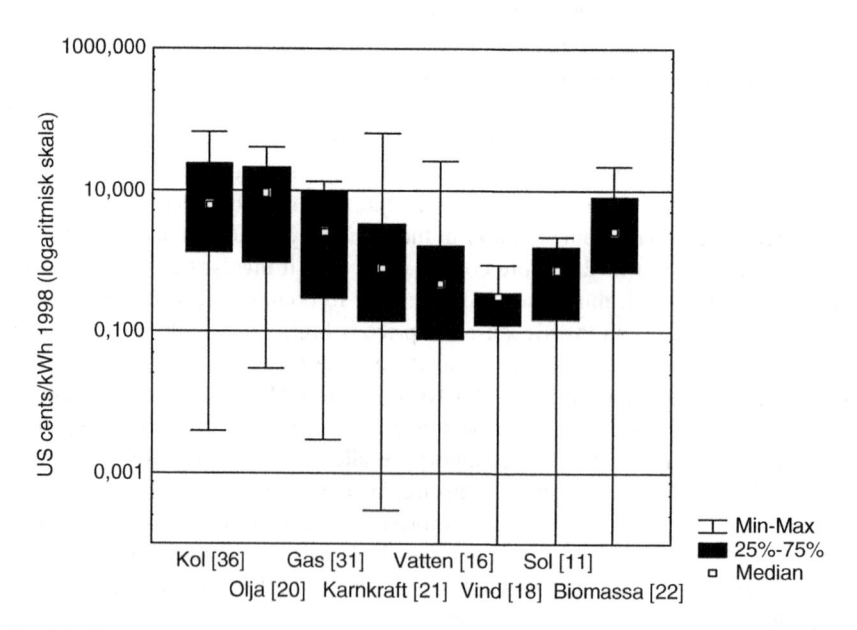

Fig. 5.5 Externalities from electricity generation. A survey of studies (The figure is taken from Michanek & Söderholm (2007). Number of studies for each source is in *brackets*. Keys: Kol = Coal, Olja = Oil, Gas = Natural Gas, Kärnkraft = Nuclear power, Vatten = Hydropower, vind = Wind, Sol = Sun, Biomassa = Biomass)

Table 5.5 Estimated emissions (kg/GWh) from a increase of production in a Danish power plant (Official data: arithmetic means of emissions from Coal powered plants in Eastern Denmark, source: Miljrapport 2008, Baggrundsrapport, Energinet.dk. Lit., review: estimates based on a small literature review)

Type of source	Emission type	Emission (kg/GWh)
Official data	NOx	911 (upper bound)
Official data 10	NMVOC	41.5
Official data	Particulates	22
Official data	CO	148
Official data	SO_2	254
Official data	CH_4	248
Official data	N_2O	8.6
Lit. review	SO2	2000
Lit. review	Arsenic	0.18
Lit. review	Cadmium	0.0025
Lit. review	Lead	0.08
Lit review	Selenium	0.0045
Lit review	HF	18
Lit. review	HCL	75

NMVOC Non Methane Volatile Organic Carbon

This can be considered an upper bound on the fossil-fuel generated emissions due to the project. Because the Danish plants are in the European carbon trading system, there will be no change in the total carbon emissions within the bubble. However, emissions of other substances will increase. Table 5.5 presents emission estimates for coal plants located in Eastern Denmark per GWh of power generation.

Data labeled 'Official data' in Table 5.5 are from studies undertaken at the Danish plants. The data labeled 'Lit. review' are to be regarded as very rough estimates, because we have no first-hand information from the Danish plants. Next, given our framework, the basic and conventional assumption is that we are concerned with the welfare of Swedes. Therefore, in a detailed assessment, we would use a dispersion matrix, showing how much of the newly generated emissions that end up on Swedish soil. However, Swedes might also be concerned with the consequences of their actions for the environment and human beings (through mortality and morbidity) abroad. We might label such concerns altruistic concerns.

We do not currently have reliable estimates for all the shadow prices in Table 5.5. We will therefore use a simpler approach in the empirical CBA. The estimate is based on EcoSenseLE V1.3, a web-based tool for estimating externality costs within the EU.[16] It provides a rough estimate of the shadow prices of increased morbidity and mortality, damage on crops and materials, and climate gases that are not covered by the European carbon trading system. We include emissions of NO_x, SO_2, particulates, NMVOC, CH_4, and N_2O according to the official data in Table 5.5; EcoSenseLE accounts for these emissions but not for the other ones in Table 5.5. The annual cost is estimated to EUR 3,039 per GWh (about SEK 30,500). This estimate is assumed to reflect the annual compensation Swedes in total request

---

[16]See http://ecoweb.ier.uni-stuttgart.de:80/ecosense_web/ecosensele_web/frame.php. This web-based tool can be used to replicate the externality costs presented in this book.

in order to be indifferent to the emissions per GWh caused by replacement power. Thus in the base case we assume that Swedes are altruists in the sense that they care about damage inflicted on others even if these people, animals, plants, and so on, live in foreign countries.

If an emission is subject to an emission tax the externality is partly or wholly internalized. Since EcoSenseLE does not account for emission taxes our approach could be interpreted as equivalent to assuming that Swedes are paternalistic altruists. That is, Swedes care about the "physical" impact of emissions ignoring that externalities might be fully or partially internalized through taxation. Even if we accept the altruism-assumption it is far from obvious whether our approach overestimates or underestimates the true damage. On the one hand, some particulates are not accounted for by the EcoSenseLE V1.3 model. On the other hand, replacement power production crowds out some other activity in the carbon permits market; recall that there are a *fixed* number of permits. In principle one would need a *global* computable general equilibrium model linked to an emissions model to be able to track down the net impact on different emission sources. Unfortunately, no such model is currently available. Still, in the the sensitivity analysis we will try to provide reasonable upper and lower bounds for the externality.

Chapter 6
Blueprint for a CBA

We have deliberately referred to this chapter as a *Blueprint*. The reason is that on the one hand we base our evaluations on solid theoretical foundations. On the other hand we lack detailed data or only have access to data from small pilot studies. Still we want to use the incomplete data set to illustrate how to undertake a modern cost-benefit analysis of projects involving environmental and recreational consequences. We present point estimates of the different items that constitute our cost-benefit analysis. This provides us with point estimates of the social profitability of Scenario 1 and Scenario 2, respectively. We then turn to a brief discussion of different kinds of sensitivity analysis and also provide an extensive sensitivity analysis where we use different distributional assumptions to illustrate possible ranges of central parameters of the cost-benefit analyses. This in turn allows us to provide bounds for the considered project's social profitability.

6.1 Point Estimates

Equipped with estimates of the different items in the cost-benefit analysis we are ready to present the results. For convenient reference the formulas used to calculate the different present value terms are collected in Appendix B.

6.1.1 SCENARIO 1

In Scenario 1, the point estimate of the present value revenue loss of the hydro-electricity plant is SEK 59 m. We estimate that present value variable cost savings amount to about SEK 3 m. Thus the present value profit loss, denoted $d\pi^F_{NPV}$ in Table 6.1, is SEK 56 m.

The point estimate of the willingness to pay for this scenario is about SEK 300 per household per year during 5 years. This is an average across the two scenarios

P.-O. Johansson and B. Kriström, *The Economics of Evaluating Water Projects*, DOI 10.1007/978-3-642-27670-5_6, © Springer-Verlag Berlin Heidelberg 2012

Table 6.1 Illustration of a
(point estimate) cost-benefit
analysis of SCENARIO 1;
million SEK

Item	Point estimate
$d\pi_{NPV}^{F}$	−56
WTP_{NPV}^{RB}	26
WTP_{NPV}^{B}	0
$-WTA_{NPV}^{E}$	−2
dT_{NPV}^{2}	0
$\sum$	−32

since we are unable to detect any significant differences between them; recall that the Contingent Valuation study is a small pilot study. To repeat, for a respondent who states an interval for his or her WTP we simply use the midpoint of the interval as our point estimate. As explained above, there are about 13,000 households in the municipality of Bollnäs. Therefore the point estimate of the aggregate present value willingness to pay of those living today amounts to SEK 18 m.[1] It might be objected that all other things equal we *overestimate* benefits by using the highest willingness to pay estimate of those listed in Table 5.4. However, we use this approach as a simple way of accounting for the fact that there might also be a small willingness to pay in neighboring municipalities.

We find it unlikely that there is a willingness to pay among those living in other parts of the country. After all, we consider small projects that are not associated with or affect any unique environmental values.[2] Therefore, the "local" present value willingness to pay, i.e. SEK 18 m, is assumed to coincide with the national one for the project under analysis. It might be added that if people living outside the local community are pure altruists, then their willingness-to-pay for the proposal is zero. This is so because in the Contingent Valuation experiment local residents pay according to their willingness-to pay, i.e. their utility levels remain unchanged. This result is proven by [101].

Future generations will also value the benefits generated by the considered project. If expected life time is 80 years, 1/80 generations or just over 160 households enter each year, assuming a constant population over time. If their WTP coincide with the one of those living today a present value of over SEK 7 m is added to the benefits[3] so that WTP_{NPV}^{RB} amounts to SEK 26 m ($18.2 + 7.5 \approx 26$). It should be noted that in principle the considered scenario is reversible in the sense that one

---

[1] Assuming that interest is compounded continuously, the present value of a stream of SEK 1 per year for 5 years is equal to SEK 4.64. It should be added that no major change has occurred in the consumer price index since the survey was undertaken.

[2] If people are willing to pay for virtually all environmental projects it might be reasonable to look for the most cost-effective way of achieving similar benefits to those provided by the scenarios considered here.

[3] $18.2(1/80)32.96 \approx 7.5$, where 18.2 is the WTP of those living today and 32.96 is the present value of a SEK per year for 150 years which is the assumed time horizon; see also Eqs. B.5 and B.6 in Appendix B.

generation might buy the water rights while another might sell them back in order to be able to hire more teachers, more nurses to the elderly care, and so on.

The effects of SCENARIO 1 on short-term variations in the water flow are probably very small. Therefore, our point estimate of WTP^B_{NPV} is zero.

Finally, replacement power is likely to cause harmful emissions of some kinds of climate gases that are not in the European carbon trading system, and other gases such as sulphur dioxide or nitrogen dioxide. Recall that the marginal electricity generator in the Nordic market typically is a fossil fuel-fired plant. Unless they are optimally taxed these emissions cause a negative externality as is discussed in Sect. 5.6, implying that those affected would need a compensation in order to be as well off as if the project did not take place. This compensation represents a cost for the project. This explains the fact that the present value compensation, denoted WTA^E_{NPV} in Table 6.1, shows up with a minus sign. In the base case we assume that Swedes take full responsibility for any damage their actions cause through emissions from fossil-fired plants that replace the production loss at Dönje. Thus it is assumed that damage outside the country's borders is valued in the same way as domestic damage.

As discussed in Sect. 5.6 we do not have enough data to estimate the cost of the emission vector in Table 5.5, so we base the estimate on EcoSenseLE V1.3, a web-based tool for estimating externality costs within the EU. The annual cost is estimated to EUR 3,039 per GWh (about SEK 30,500). This yields a present value cost equal to SEK 4 m if the emissions continue "forever". However, it seems likely that emission restrictions will become tighter over time. We account for this fact by assuming that emissions to vanish after 20 years. Therefore the point estimate of WTA^E_{NPV} is SEK 2 m. Since this estimate covers some but not all types of emissions, ceteris paribus it underestimates the true cost. On the other hand, it overestimates the true cost if in the "German" price forecast scenario coal-fired plants are not replacing the production loss at Dönje. For example, the replacement power might instead be provided by pumped-storage plants. The same outcome occurs if the plants crowded out by the Danish plant in the market for carbon emission permits reduce their emissions of other harmful substances. However, there is no European or Global computable general equilibrium model linking production functions to "emission functions". Therefore we are unable to estimate the impact of the scenario on *global* general equilibrium emissions of different particulates. In the sensitivity analysis we will try to account for this uncertainty with respect to the magnitude of the emissions externality.

Since we assume that demand for electricity as well as supply of labor remains unchanged the tax term dT^2_{NPV} is equal to zero. Adding the different items in Table 6.1 we arrive a a point estimate of the social profitability of SCENARIO 1. According to this estimate SCENARIO 1 causes a loss to society of about SEK 30 m.

6.1.2 SCENARIO 2

In SCENARIO 2 the point estimate of the present value loss of profits is SEK 217 m; see Table 6.2. Thus the loss is almost four times as large as in SCENARIO

Table 6.2 Illustration of a
(point estimate) cost-benefit
analysis of SCENARIO 2;
million SEK

Item	Point estimate
$d\pi_{NPV}^{F}$	−217
WTP_{NPV}^{RB}	26
WTP_{NPV}^{B}	0
$-WTA_{NPV}^{E}$	−7
dT_{NPV}^{2}	0
$\sum$	−198

1. Still, our small Contingent Valuation study was unable to detect any significant difference in willingness to pay for the two scenarios. Thus WTP_{NPV}^{RB} amounts to SEK 26 m also in this scenario. It seems unlikely that people living outside the municipality of Bollnäs and its vicinity attribute any significant value to the small change that SCENARIO 2 represents. Therefore, WTP_{NPV}^{RB} is assumed to represent society's present value willingness to pay for the project. This scenario might have a larger impact on the water flow than SCENARIO 1 since the involved amount of water is larger. However, the *impact* on short term *variations* in the flow is probably still small and difficult to detect by the typical citizen. For this reason WTP_{NPV}^{B} is assumed to be zero also for SCENARIO 2. Finally, we assume that electricity lost at Dönje is replaced by increased generation by a fossil fuel-fired plant. The present value cost of the associated damages caused by emissions is estimated to SEK 7 m. The argument is the same as in SCENARIO 1, that Swedes are concerned about damages inflicted by Sweden's electricity consumption independently of where the electricity is generated and the damage occur. Summing the different items we arrive at a point estimate of the social profitability of SCENARIO 2. As is seen from Table 6.2 the project causes a loss which is estimated to about SEK 200 m.

6.2 Sensitivity Analysis

We now turn to a sensitivity analysis. Before illustrating the sensitivity of the results to changes in central parameters we briefly discuss different approaches to sensitivity analysis. In particular we undertake several stochastic sensitivity analysis. The aim is to illustrate how different tools can be used to show the decision-maker the sensitivity of the outcome of a cost-benefit analysis to distributional assumptions.

6.2.1 Some Different Approaches to Sensitivity Analysis

A sensitivity analysis shows if the results are sensitive to substantial but plausible variations in crucial parameters. Hence it judges the robustness of the conclusions of the CBA. A one-way sensitivity analysis varies one parameter at a time. If the best estimate is used in the base case CBA, we might be able to locate extreme values in a plausible range of the parameter and use these in the sensitivity analysis. Alternatively, it might be possible to construct a confidence interval for the parameter. If one is unable to find a value or range for the parameter one could perform a threshold analysis. In such an analysis the parameter is assigned a value such that the outcome of the CBA takes on a chosen value, for example, shows a zero result.

To undertake a single univariate sensitivity analysis does not make much sense. In the more reasonable case where multiple univariate sensitivity analyses are undertaken they are often presented in a Tornado diagram.[4] In such a diagram the parameters are ordered according to their impact on the result from widest range to most narrow range.

Sometimes one must undertake multivariate sensitivity analyses. For example, parameters might be correlated so that their total impact is larger or smaller than the sum of their impact according to univariate analyses. In the simplest case, one considers two-way sensitivity analysis where two parameters are varied at a time. A typical approach is to construct a diagram in which the two parameters are varied so as to keep the result unchanged, i.e. one construct a type of indifference curves or isoquants where each curve keeps the result at a particular level. However, it is unlikely that both parameters take on their extreme values together. Therefore, one might try to find a set of parameter values that provide a likely upper bound and a lower bound, respectively, for the outcome.

There are also probabilistic sensitivity analyses where probability distributions are assumed for different parameters. This approach has become popular in cost-effectiveness analysis.[5] For example, mean costs are often assumed to follow either a log-normal or a gamma distribution. Monte-Carlo techniques are then used to generate the results. Sometimes, Cost-Effectiveness Acceptability Curves are constructed (based on parametric or non-parametric methods). Such curves provide the probability that the difference between benefits and costs is positive. Typically, benefits are defined as the average WTP for a quality-adjusted life year.[6] Hence this approach is a kind of aggregated or average CBA rather than a conventional cost-effectiveness analysis, where the focus is on the ratio between costs and an effect measure, often quality-adjusted life years. For more on different approaches to sensitivity analyses in cost-effectiveness analysis, the reader is referred to [24]. We will suggest some different approaches to stochastic sensitivity analysis in cost-benefit analysis later on in this study and also introduce some tools and approaches we believe are novel in the cost-benefit literature.

---

[4]For an illustration, see http://www.tushar-mehta.com/excel/software/tornado/decopiled_help/tornado.htm.

[5]The reader is referred to [87] for an interesting attempt to compare different methods to estimating confidence intervals for WTP data.

[6]Typically health quality runs from zero (death) to unity (perfect health).

6.2.2 Illustrations

We are not able to undertake a sophisticated multivariate sensitivity analysis in this book. We only present very simple univariate analyses in order to illustrate the sensitivity of the outcome to arbitrary changes in base case parameter values; throughout we report deviations from the point estimates reported in Tables 6.1 and 6.2.

As is seen from Table 6.3 the price trajectories have a quite considerable impact on the outcome of the cost-benefit analysis. This wide range is due to the fact that we work with two quite different equilibrium trajectories for the spot price of electricity. This is done in order to locate a reasonable lower bound and upper bound, respectively, for the long-run development of the spot price. The symmetry is simply due to the fact that the point estimates in the base case evaluations are the averages of the two trajectories. Note that in Table 6.3 a minus sign means that the societal loss associated with the considered change in water use increases while a positive sign indicates that the loss is reduced.

Fortum has provided us with an estimate of the revenue loss using their optimization model for the operation of power plants. They assume an average winter price of slightly above SEK 500 per MWh and an average summer price of around SEK 400 per MWh; these estimates were made in late September 2008. Fortum estimate that the annual loss amounts to SEK 1.5–2 m in SCENARIO 1 and to at least SEK 7 m in SCENARIO 2. In comparison, the simple way of estimating the loss of production that is used in this book, see Sect. 4.1, result in revenue losses of SEK 1.85 m and SEK 7.2 m, respectively, when the spot price is SEK 500 per MWh. Unfortunately, Fortum made no price forecast so it is difficult to say whether their calculations would result in higher or lower revenue losses than those reported in Tables 6.1 and 6.2. However, our starting-point price in the base case evaluation is SEK 350 per MWh, i.e. considerably below the (2008) prices used by Fortum. Still, we believe that our starting-point price is reasonable since spot prices in 2008 were among the highest ever recorded due to extreme market conditions; see Sect. 4.2.1 for details.

In the base case it has been assumed that future investment costs at the Dönje plant are unaffected by the considered small changes in production. However, the term I^τ in Table 6.3 shows how the CBA is affected if future investment costs

Table 6.3 A rough sensitivity analysis. Results indicate the change in the outcome of the CBA in SEK million in comparison to the point estimates	Item	SCENARIO 1	SCENARIO 2
	p^s-trajectory, lower bound	16	62
	p^s-trajectory, upper bound	-16	-62
	I^τ	6	24
	$r = 0.06$	25	112
	$r = 0.015$	-44	-189
	WTP^{RB}_{NPV}, upper bound	8	8
	WTP^{RB}_{NPV}, lower bound	-8	-8
	$-WTA^{E}_{NPV}$, upper bound	-2	-8
	$-WTA^{E}_{NPV}$, lower bound	2	7

are proportional to the water flow devoted to electricity generation. In this case it becomes slightly cheaper to let water bypass since society can avoid some future investments.

It might be noted that if the remaining life of the existing plant is halved from 40 to 20 years, the societal losses caused by the considered projects/scenarios *increases* slightly (a result not shown in the table). The reason is the fact that in the base case we assume that future investment costs are unaffected by the considered small projects and that the operating costs of future plants are virtually zero. On the other hand, by assumption the existing plant has a positive marginal operating cost. Hence the sooner the existing plant is replaced the lower is the present value of the resources that are saved due to the considered projects/scenarios. In other words, if the existing plant is scrapped immediately, no resources would be released for other uses since all future fixed and variable production costs remain unchanged.

To the best of our knowledge, Dönje is not eligible for energy certificates. However, when the plant is refurbished next time one cannot rule out that it is "redesigned" in such a way that it becomes qualified (insofar as the rules in the future permit this). If so, the loss of revenue will be considerably higher than we have estimated. This is so because the certificate price is quite high and seems to be increasing over time; during 2008 the average price was about SEK 250 per MWh. This price corresponds to almost SEK 1 m (SEK 3.6 m) per year for SCENARIO 1 (SCENARIO 2). We refrain from any attempt to estimate present values since we lack sufficient information for undertaking such an exercise.

From Table 6.3 it is obvious that the social discount rate has a dramatic impact on the outcome; in the table we only account for the impact on the present value loss of profits. If the discount rate is halved, the cost of implementing the considered projects/scenarios increases sharply. If the discount rate is doubled from 3% to 6% the losses are sharply reduced. In fact, if the discount rate is sufficiently high, the sign of the CBA is reversed even in our high electricity price scenario. This is due to the fact that the present value of large profits losses in a distant future goes to zero as the discount rate goes to infinity. The willingness to pay of the current generation, on the other hand, is not very sensitive to the choice of discount rate since it refers to the present value of annual payments during 5 years; it increases (decreases) by about SEK 1 m if the discount rate is halved (doubled). However, if electricity prices continue to grow, then at some point in time, say time t, the present value of the future profit loss will be so large that it is profitable for Fortum to repurchase the water use rights. Thus at time t we would "switch back" to using the water for electricity generation. On the other hand, WTP_{NPV}^{RB} might grow over time. Such a growth could be due to growing incomes or a shift in preferences so that a referendum at time t results in a higher WTP_{NPV}^{RB} than a referendum today. Such changes or a dramatic fall in electricity prices might make the project(s) worthwhile at some future point in time. In Sect. 6.2.4 we briefly discuss a possible optimal implementation date. However, in this section our objective is more modest, namely to assess the social profitability if the projects are implemented *today*.

The referendum question of our Contingent Valuation study did allow respondents with a positive willingness to pay to state either a maximum WTP or an

interval indicating a range covering their "true" WTP. We have used these intervals
in the sensitivity analysis. The upper bound is interpreted as the maximal WTP while
the lower bound is interpreted as the minimum WTP. Respondents stating a single
number for their WTP are assumed to thereby reporting the same number for their
best guess, upper bound and lower bound, respectively, of their WTP.[7] Given this
approach the upper bound for the present value willingness to pay is SEK 8 m higher
than the base case present value WTP. The lower bound is SEK 8 m below the base
case number. Note that we once again consider the average over SCENARIO 1 and
SCENARIO 2 since we are unable to detect any significant differences.

In establishing point estimates the damage caused by replacement power was
assumed to vanish after 20 years. If the damage remains throughout the considered
time horizon (150 years) the cost increases by SEK 2 m in SCENARIO 1 and SEK
8 m in SCENARIO 2. Assume instead that Swedes act as "nationalistic" egoists in
the sense that any damage inflicted on living species (including human beings) or
the environment *abroad* is ignored (or that net emissions are unaffected for one
reason or the other). If all damage occurs abroad or is unaffected by the considered
reregulation this amounts to setting $WTA_{NPV}^E = 0$ which improves the results to
some extent as is seen from the final row of Table 6.3.

6.2.3 Demand Changes

According to Eq. 3.4 demand for electricity might change due to small price changes
caused by the considered project. Even if such demand changes might seem unlikely
they cannot be ruled out. For illustrative purposes let us assume a Cobb-Douglas
demand function $x^d = (18975 \times 10^7)/p^d$. This specification yields a demand equal
to 150 terrawatt hours (TWh) or 150×10^6 MWh, which is a typical annual demand
level for Sweden, when the *consumer* price p^d, see the discussion following Eq. 6.1
below, is SEK 1265 per MWh.[8] Then a price increase of around SEK 0.03 per MWh
is sufficient to reduce demand by 3.7 GWh, which is the same amount as the loss
of production in SCENARIO 1. Intuitively, the shift to the left of the "supply ladder"
might mean that a marginally more costly power plant is used a little more during
the year. This pushes the average price slightly upwards and so reduces demand.
One can hardly observe or measure these small changes unless a detailed numerical
supply and demand model of the market is available, but they could still be present.
Even if we replace the Cobb-Douglas demand function, whose price elasticity is -1,
with functions generating more realistic price elasticities, say -0.3 to -0.4, a price
increase of no more than around SEK 0.1 per MWh is sufficient to decrease demand

---

[7]The base case average WTP is SEK 301 per household for 5 years, while the upper (lower) bound
is SEK 399 (SEK 203).

[8]We assume that the spot price is SEK 480 per MWh since this is our certainty equivalent price;
see the discussion in Sect. 4.2.2.

by 3.7 GWh. Even if we allocate the entire demand reduction to the household sector, which accounts for 20–25% of the total electricity consumption, a price increase of around SEK 0.4 per MWh seems to be be sufficient to accomplish the considered demand reduction with a price elasticity of −0.3.

Therefore, we supply a rough estimate of the magnitude of the value of such changes. Drawing on Eqs. 3.4 and 3.5 and for computational simplicity assuming that only households living in flats in some parts of the country reduce their consumption we arrive at the following estimate of the present value loss of governmental tax revenue,

$$dT_{NPV}^{x^d} = dx^d \int_0^\infty [282 + [480 + 282 + 50 + 200] \times 0.25] e^{-rt} dt, \qquad (6.1)$$

where dx^d is the change in final demand for electricity caused by the considered project, the unit tax on households' electricity consumption is SEK 282 per MWh as of fall 2010, the spot price is assumed to be SEK 480, the before VAT price a final consumer pays for energy certificates is assumed to be SEK 50 per MWh, the variable transmission price (including any cost for regulating services) is assumed to be SEK 200 per MWh, and VAT is currently 25%. Vattenfall, E.ON and Fortum, the three largest players on the Swedish electricity market, have variable transmission prices of SEK 174–240 per MWh for consumers living in flats[9] (as of fall 2010). Here we assume that such consumers are those who following small changes in prices adjust their consumption and that the average consumer faces a variable transmission price of SEK 200 per MWh. According to Eq. 6.1 we assume that all price and tax components except the spot price remain constant over time. This assumption probably causes an underestimation of the true loss since one would expect at least the unit tax on electricity to be increased over time.

This tax term has a huge impact on the outcome of the CBA. In SCENARIO 1 the social *gross* loss increases by some SEK 65 m and in SCENARIO 2 the gross loss increases by SEK 250 m. This is due to the fact that electricity consumption in Sweden is subject to very high taxation. The numbers illustrate how the outcome is changed if electricity demand falls at the same rate as production at the Dönje plant. It should be stressed that in this demand reduction case there is no replacement of the electricity lost at Dönje. Therefore the positive amount WTA_{NPV}^E should be added to the negative amount $dT_{NPV}^{x^d}$.

We probably overestimate the tax income loss because consumers might increase demand for other taxed goods than electricity (or possibly spend the same amount of money on electricity as initially, depending on the price elasticity of electricity demand), that's why we above speak of a "*gross* social loss". If the net impact

---

[9]E.ON Stockholm småförbrukarprislista (SEK 239.2 per MWh), Fortum enkeltariff, Stockholm (SEK 194 per MWh), Vattenfall Söder enkeltariff E4 (SEK 174 per MWh). These prices are available on the home pages of E.ON, Fortum, and Vattenfall, respectively (as of fall 2010).

is restricted to the electricity tax in Eq. 6.1 the loss is around SEK 35 m and SEK 135 m, respectively, for the two scenarios under consideration. However there is a demand-depressing loss of income through the compensation paid to the hydropower firm in order to induce it to divert water from electricity generation so these numbers might underestimate the true loss.

According to Eq. 3.5 and A.19 in Appendix A there are also tax wedges on the cost side. If the considered reduction of electricity generation causes aggregate use of taxed inputs to change, then we have a similar term as the one considered in Eq. 6.1. Both the marginal cost of current production (exclusive of the water value) and future replacement investments might be associated with such tax wedges. However, these effects are likely minor since the marginal production cost and the present value of future investments are small. Therefore they are ignored in this sensitivity analysis.

6.2.4 A Simple Krutilla-Fisher Robustness Check

Let us also present a kind of simple robustness check. Instead of basing the analysis on forecasts of prices and other parameters we now as far as possible base the cost-benefit analysis on current or known parameters. Unfortunately, there is no unique or time-invariant spot price for electricity as is evident from Fig. 4.2. However, according to the linear trend in that figure the "normal" price at the end of 2010 is around SEK 440 per MWh, noting that the average spot price (system price) for the "credit crunch" year 2009 was in excess of SEK 370 per MWh. Assuming that the price stays at this level "forever" yields a present value loss of profits of around SEK 51 m for SCENARIO 1 and SEK 196 m for SCENARIO 2, respectively. Add the quite small externality cost of replacement electricity and this robustness check indicates that *presently* it is not socially profitable to undertake any of the two scenarios since the present value benefits amounts to around SEK 25 m for both scenarios, assuming here that demand for electricity remains unchanged.

Reference [121], presented in Chap. 2 find it likely that the benefits of development of a natural resource decrease over time. One reason is technological change over time. The benefits of preservation, in contrast, are likely to increase over time. One reason is that technological change can hardly produce close substitutes to environmental resources, whereas increasing incomes tend to increase WTP for environmental resources as long as they are normal, as is typically assumed. This so-called Krutilla-Fisher model has attracted much interest in the literature and suggests that our scenarios might be profitable in the long run.

However, there is also the question of *when* to optimally implement a reversible project; as explained in Chap. 2 there is a pure postponement value, denoted *PPV* in Eq. 2.2 which provides the Dixit-Pindyck real option value, even in a deterministic setting. Given current prices, incomes, and preferences, as stated above, "today" it seems socially unprofitable to implement any of the considered scenarios. In order to reverse outcome of the CBA either we must await an increasing WTP for the scenarios over time, for example, due to increasing incomes, or a falling spot price of electricity. To illustrate, assume that the *total* WTP for SCENARIO 1 increases by 1.8% year after year, i.e. at the same rate as GDP per capita over the last four decades, see Sect. 4.3, all other things being equal. Then it would be optimal to postpone the project around 45 years; then, its present value benefits would match its present value costs (and if benefits continue to grow beyond year 45, as seen from today, it will be profitable). Given the same assumptions, SCENARIO 2 should be postponed almost 120 years. Alternatively, one might drastically change the assumptions underlying the analysis. However, the conventional approach to investment evaluation is to look for lower bounds for the benefits and upper bounds for the costs rather than the opposite. If a project passes such a strict robustness or stress test it is likely to be profitable. Currently, our scenarios do not seem to pass such tests.

6.2.5 A Stochastic Sensitivity Analysis

Some of the items of the cost-benefit analyses are stochastic. We have assumed that the spot price at each point in time is an iid uniform random variable on the interval (p_t^{slb}, p_t^{sub}), where p_t^{slb} (p_t^{sub}) denotes the lower bound (upper bound) estimate of the spot price at time t. This is the simplest possible continuous random process and means that the point estimate of the spot price at time t is equal to $(p_t^{slb} + p_t^{sub})/2$. In turn, this assumption generates a lower and an upper bound for the present value loss of profits; the lower bound is SEK 40 m and the upper bound is SEK 72 m, concentrating on SCENARIO 1 here. Moreover, parametric estimations of the willingness to pay for the considered projects involve distributional assumptions. We will briefly discuss two bivariate distributions for these two random variables, loss of profits and WTP; see Eqs. B.10 and B.11 in Appendix B. The first variation assumes that the present value loss of profits follows a uniform distribution on the interval $[-72, -40]$. The *aggregate* present value WTP for the considered scenario, i.e. WTP_{NPV}^{RB}, is assumed to be normally distributed with mean 26 and variance 6.25. These latter assumptions ensure that virtually all probability mass is in the interval $[17, 34]$, the assumed lower and upper bounds respectively for the aggregate present value WTP. A second variation assumes a normal distribution also for the loss of profits with a mean equal to 56, as in Table 6.1, and a variance equal to 25. These assumptions ensure that virtually all probability mass is in the assumed interval for the loss of profits.[10]

We assume that the two random variables are independent. It is hard to see that there should be any dependence of the Nordic spot prices on WTP and vice versa. The resulting joint probability density functions are shown in Fig. 6.1. The present

---

[10] A possible refinement would be to truncate (and rescale) the distribution so that all probability mass is in the interval $[-72, -40]$.

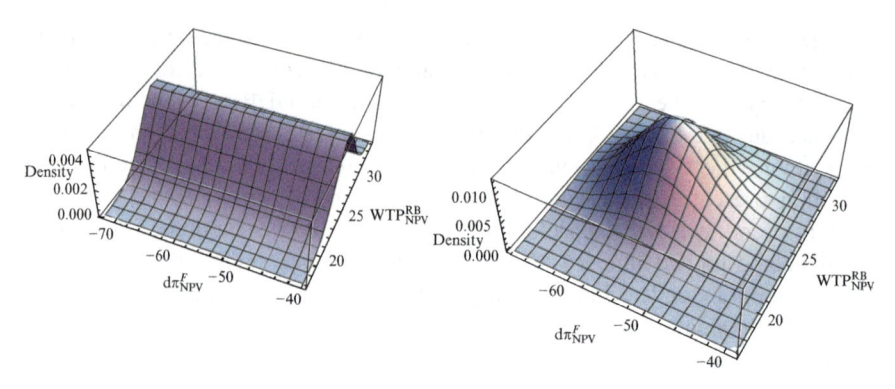

Fig. 6.1 Probability density functions with WTP_{NPV}^{RB} and $d\,\pi_{NPV}^{F}$ in million SEK

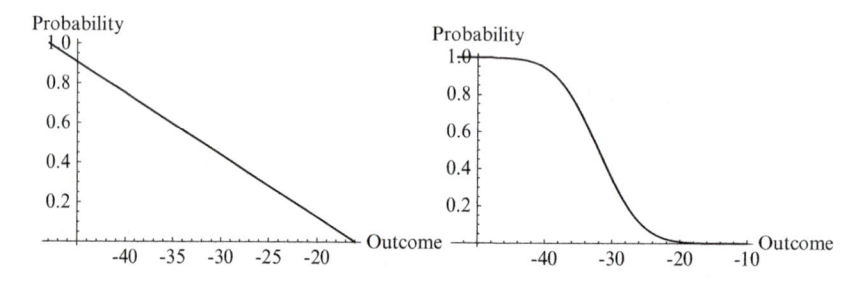

Fig. 6.2 Illustration of cost-benefit acceptability curves with outcome in million SEK

value loss of profits is denoted $d\,\pi_{NPV}^{F}$ while the aggregate present value willingness to pay is denoted WTP_{NPV}^{RB} and the density is measured along the vertical axis. The left function in the figure refers to the case with a uniform distribution for the loss of profits while the right function refers to the case where both variables follow normal distributions. It is apparent from the figure that the outcome is in negative territory. In any case the figure provides one and useful way of illustrating the outcome of a cost-benefit analysis when some variables are assumed to follow particular statistical distributions.

Another illustration is provided by Fig. 6.2, which also refers to Scenario 1. The curves depicted in the figure might be termed *cost-benefit acceptability curves* since they yield the probability that the social profitability *exceeds*, say, SEK x million; the left (right) panel refers to a uniform (normal) distribution for the profits loss.[11] We believe a decision maker will find such curves more informative and relevant than curves (i.e. cumulative distribution functions) yielding the probability that the outcome is SEK x million or less. The curves in the figure have been constructed by assuming that the aggregate WTP for Scenario 1 is SEK 26 m, and that the emission externality amounts to SEK 2 m. Thus these variables are

---

[11]The associated survivor functions are stated in Eqs. B.12 and B.13 in Appendix B.

stripped of any uncertainty. All uncertainty has been allocated to the loss of profits. In practice, we simply shift the probability density function to the right by an amount equal to SEK 24 m, i.e. undertake a spread-preserving shift in the mean of the profits loss. The curves in Fig. 6.2 refer to survivor functions (i.e. one minus cumulative distribution functions). Therefore, according to both approaches the probability is more or less nil that the loss is smaller than SEK 20 m and it is almost certain that the loss is no more than SEK 50 m. This is possibly a useful way of illustrating the likely outcome of a project in an uncertain world. Setting WTP equal to its upper (lower) bound value would shift the acceptability curves to the the right (left).

In the more general case both the loss of profits and the WTP are considered to be random. The associated general expression for the cost-benefit acceptability curve is stated in Eq. B.14. Assuming normally distributed and independent variables the joint survivor function is close to the survivor function in the right panel of Fig. 6.2; it runs below (above) for probabilities exceeding (falling short of) 0.5. Thus allowing both the change in profits and the WTP to be random does not change the picture of the scenario's likely social profitability in any dramatic way. Allowing the externality cost to be a normally distributed variable with mean -2 and variance 0.25 so that the almost all of the mass is in the interval $(-4,0)$ increases the total variance but does not much affect the cost-benefit acceptability curve. However, this approach becomes quite complex if many variables are random and possibly correlated. Therefore, in the next section we use a simulation approach that is more easily implemented.

Still another and easily implemented approach is to assume that we "draw" extreme outcomes from distributions for the spot price trajectory, the annual WTP for the considered scenario, the discount rate, and the annual externality cost; for simplicity the distributions are assumed to be independent and we once again ignore the possibility that other factors than those listed in Table 6.3 might be stochastic. At one extreme, corresponding to the lower bound for the project's profitability, we happen to draw a spot price that develops according to the scenario where the Nord Pool market and the EEX "merge" in 20 years, the annual WTP for SCENARIO 1 is SEK 203 per household, the social discount rate is equal to 1.5%, and the externality cost of replacement power remains forever. This combination of draws results in a present value loss of SEK 124 m which is the *lower* bound for SCENARIO 1's social profitability. The other extreme occurs if the draws from the four different distributions result in a spot price which remains constant at SEK 350, an annual WTP per household of SEK 399, a discount rate of 6%, and a zero externality cost of replacement power. This combination yields a present value gain of SEK 7 m which is the *upper* bound for the scenario's social profitability.

There are an infinite number of other draws resulting in outcomes in the interval $(-124,7)$, tracing out the probability density function (or probability mass function if we think in terms of a discrete probability distribution). If the mean still is SEK -32 m we know that the distribution must be skewed, although we have not introduced sufficient assumptions to determine its shape. One distribution that fits the "data" reasonably well and is able to generate a mean equal to -32 is the Gumbel (minimum) or extreme value type I distribution; see [70] or [35] for details.

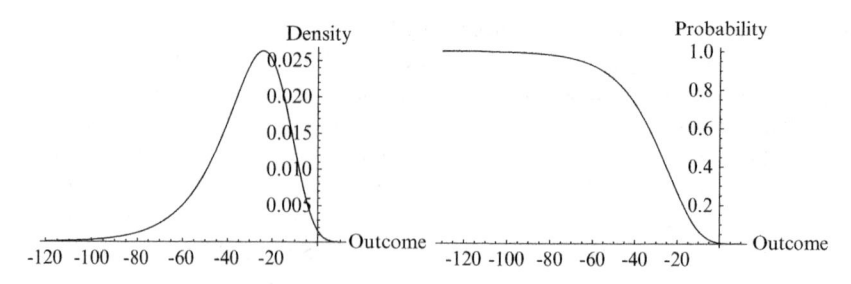

Fig. 6.3 Illustration of probability density function and cost-benefit acceptability curve with outcome in million SEK assuming a Gumbel distribution

For illustrative purposes we set its location parameter (mode) equal to -24 and its scale parameter equal to 14 to obtain the probability density function pictured in the left panel of Fig. 6.3. Almost all probability mass falls in the interval $[-124, 7]$.[12] The associated survivor function or cost-benefit acceptability curve is pictured in the right panel and suggests that it is highly unlikely that SCENARIO 1 is socially profitable. The mean, denoted m, is equal to $m = -24 - 14 \cdot 0.5772 = -32$, where 0.5772 is the rounded-off Euler-Mascheroni constant. For further technical details the reader is referred to Eqs. B.15 and B.16 in Appendix B.

A *triangular* distribution might seem to be an interesting alternative to the unbounded Gumbel distribution since it is defined on a finite range, e.g. $[-124, 7]$ as in SCENARIO 1; see Eq. B.18 in Appendix B for the probability density function of a triangular distribution. However, the mean estimate of SCENARIO 1's profitability, i.e. -32, is too high relative to the support for the triangular distribution to be applicable. Setting the mode at (or for technical reasons just below) the right boundary of the support, i.e. 7, yields the maximal mean $(= -36.7)$ for the triangular distribution since the mean is calculated as $(a + b + c)/3$, where a (b) is the left (right) boundary of the support, and c is the mode. Thus on the range $[-124, 7]$ the mean for the triangular distribution is everywhere lower than the point estimate of SCENARIO 1's social profitability, i.e. -32. We interpret this fact as indicating that the actual distribution is too left skewed to be approximated by a triangular distribution. This is also true for the more general *trapezoidal* distribution, see Eq. B.19 in Appendix B, which generates the uniform and triangular distributions as special cases.[13] The trapezoidal probability density function viewed from left to right has a growth stage, a stage of stability, and a stage of decay.

---

[12]Loosely speaking, setting the location and scale parameters at lower (higher) values results in a density function that is too narrow (wide) given that the mean is kept unchanged. A further refinement would be to truncate the distribution so that all probability mass is contained in the considered interval.

[13]The reader is referred to [44] for a detailed treatment of the properties of the generalized trapezoidal distribution.

Given support $[-124, 7]$ the highest mean for the trapezoidal distribution is obtained for the triangular variation with the mode equal to 7 as above, so the mean is at most -36.7. There are generalized trapezoidal distributions where the growth and decay may exhibit a nonlinear behavior; see [44] for details. It is also available in the free statistical software R as Package 'trapezoid'. Similarly there are generalized discrete triangular distributions; see, for example, [113] for details. Due to lack of information with respect to the shape of the actual distribution we have abstained from introducing generalized variations in our sensitivity analysis. However, the generalized distributions are potentially very useful tools in empirical sensitivity analysis since they are compatible with almost any conceivable probability density (or mass) function, as is illustrated by Fig. 4 in [44].

6.2.6 A Stochastic Sensitivity Analysis Based on Simulation Techniques

A straightforward approach to shedding some potentially interesting light on the uncertainties is to use what is sometimes called Monte-Carlo methods, or "systematic sensitivity analysis" (see [79]). Thus, rather than using a number of different parameter values (typically representing extreme outcomes), we draw values of key parameters from a distribution and then calculate net present value, given the drawn numbers. By repeating this process a large number of times, we obtain a distribution of net present values, representing how sensitive the project is towards stochastically perturbing the key parameters. To implement the approach we use the formulas in Appendix B.[14]

We let the interest rate, the price forecast, household annual WTP and the number of years that the negative externality is "alive", be described by distributions. Specifically, we draw the interest rate from a truncated normal with mean 3%, standard deviation 1% and lower (upper) limit 1.5% (6.0%). There is no particular reason to use a truncated normal, many other distributions will be useful. We use the truncated normal to ensure that the random variables is within the stated limits. The price forecast (i.e. the constant real growth rate of the electricity price) is taken from a uniform distribution on the interval $[0, 0.05]$. Here again one could explore different distributions, this choice simply reflects that we do not have any particular information that would motivate any other stochastic assumption. We assume that the price grows (with a constant rate) over the first 20 years, after which it is constant. With these assumptions, we obtain for each draw a particular value of the profit change, the change in investment costs, WTP and WTA and thus obtain a net value of the project. Each draw is so to speak "one world", the drawn parameter configuration, in which the whole project lives. Each particular value depends in

---

[14]To implement this in R, libraries `Ryacas`, `triangle` and `msm` are convenient. These libraries are available for automatic download from CRAN.

a complicated way on the stochastic assumptions made on the key parameters and the particular functional forms used. Clearly, this is a very simple set-up and there are, indeed, many ways to make a more sophisticated simulation. For example, we can let the initial spot-price on electricity be a random variable, the duration of the project might be stochastic and so on and so forth. In addition, we can use other and more general statistical distributions. The basic purpose here is to illustrate in a simple way how uncertainty can be addressed in project evaluation. In this blueprint we prefer simple and direct approaches and rather try to make the point that sensitivity analyses should always be undertaken in CBA.

Before turning to the results, let us comment briefly on the assumptions made regarding WTP and WTA in this simulation. WTP is drawn from a triangular distribution with limits SEK (203,399) and mode 300.83. This choice is influenced by the analysis in [188]. We could also have used the estimated Weibull from the self-selected intervals, but in this way we also keep a closer link to the other sensitivity analyses above. At any rate, WTP depends in the simulation on two factors, the interest rate and household annual WTP. The longevity of the externality caused by replacement power is either 0, 20 or 150 years (the assumed time horizon of the project) with given probabilities.

In each given run we compute the social profitability of the project, given the values of four random variables. By repeating this process we thus obtain a distribution of possible outcomes. Observe that this specification replicates (although not exactly) the upper and lower bounds described in the previous section.

Figures 6.4 and 6.5 summarizes one simulation. Figure 6.4 displays the outcome for the present value of the loss in profit and the present value WTP for SCENARIO 1. There seems to be a slight skewness in the outcomes. In Fig. 6.5 the empirical version of the cost-acceptability curve curve, based on the simulation, is shown. There are several ways to approximate such a curve from data. Here we have simply used the empirical cumulative density functions (using the default settings in the

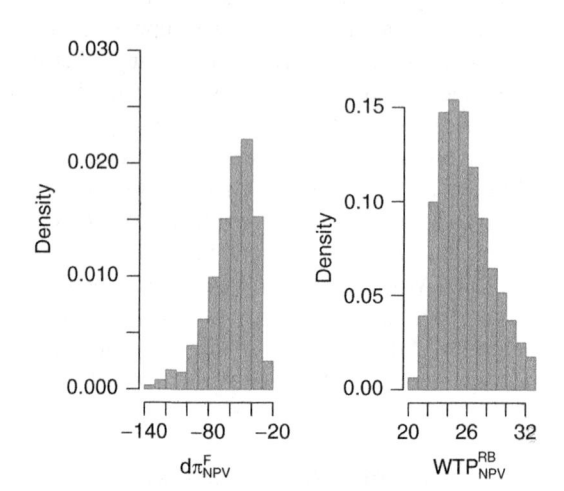

Fig. 6.4 A systematic sensitivity analysis in million SEK based on 10,000 replications

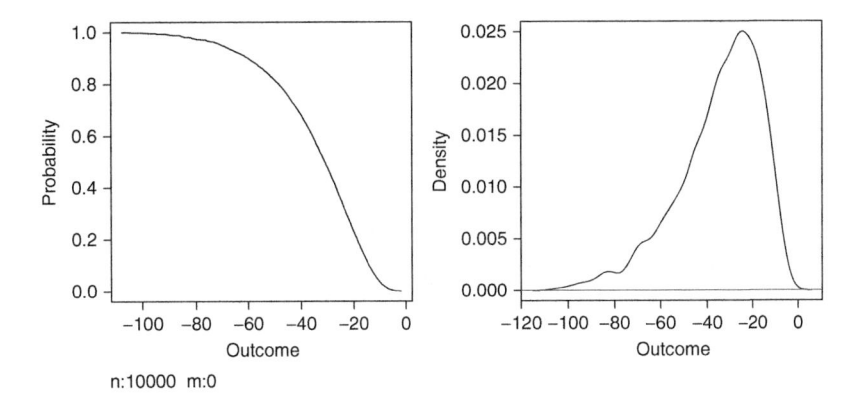

Fig. 6.5 Illustration of a cost-benefit acceptability curve and the associated probability density function from the systematic sensitivity analysis, with outcome in million SEK

command ECDF in R). The right-hand side of this figure uses the default settings of the density command in R. Evidently, these simulations suggest that there is very little chance for the project to be socially profitable.

6.2.7 Risk Aversion

There is also the question how risk aversion enters the picture. For example, we have two different price forecasts assumed to occur with probability 0.5 each. There is also uncertainty with respect to the benefits of our scenarios. Using probabilities of different outcomes to weigh them one can calculate an expected outcome of undertaking either of our two scenarios. This is basically the approach behind the results reported in Sect. 6.1. The question is how such expected outcomes are related to receiving or paying an amount with certainty to escape the risk. Consider Fig. 6.6. A risk averse individual or society faces an uncertain outcome of an investment. Either he gains a positive income G or loses an income equal to L, normalized so that the origin represents the ex ante situation.[15] The expected loss is denoted A yielding an expected utility equal to U. This risk averse individual would accept taking a larger certain loss or certainty equivalent than A because the pain of the bad outcome is so high. In effect, a *certain* loss equal to B yields the same utility, U, as the "game". Thus the individual is willing to give up an amount equal to $A - B$ to get rid of the investment's uncertainty. In this sense, and all other things equal, our risk-neutral approach might seem to underestimate the true social loss, at least as long the expected outcome is in negative territory. Recall that our investment is to divert water from electricity generation causing some benefits but also causing uncertain

---

[15]The argument remains unchanged if we assume that both outcomes result in losses.

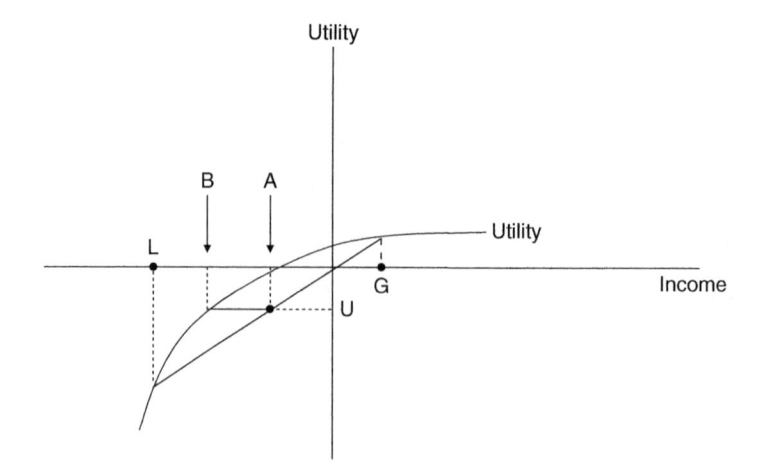

Fig. 6.6 Illustration of a certainty equivalent

costs. However, as the scale of the project is reduced the underestimation of the loss
is reduced too. For very small projects, like those considered in this study, the underestimation is close to zero since the points G and L virtually coincide in Fig. 6.6.

6.3 On the Choice of Discount Rate and the Possibility of a Local Deal

A final issue deserves discussion. Firms often use much higher discount rates then
the social rate used in this study. In the considered case this fact means that Fortum
would be willing to sell water according to one or the other scenario for a much
lower present value price than the point estimate cost-benefit analysis indicate. This
might suggest that we overestimate the social cost of the considered projects or
scenarios. In this section we will briefly address the issue of the social discount rate
as measured from the capital side and the issue what it means from a societal point
of view if a group, say the local citizen of Bollnäs, are able to "bribe" Fortum to
divert water from electricity generation.

6.3.1 On the Marginal Cost of Capital

Financial markets are often dominated by a few large firms. It seems likely that these
large players have and exploit market power.[16] If so, they earn a profit on marginal

---

[16]This section draws on [104].

transactions, i.e. there is a markup on their own marginal costs. At least a part of the difference between discount rates used by private firms and the societal one used here might be explained by this kind of market imperfections. Thus any present value loss of profits of financial agents should be added to the firm's estimate of its present value loss if water is redirected from electricity production.

A recent study analyzes market power of Swedish banks. It estimates the Lerner index for Swedish banks to about 22% for the period 1996–2002; see [161] for details. In other words, the mark-up (the difference between price and marginal cost) over price ratio is about 0.22. The same study estimates a Translog cost function for the banking industry with three variable inputs (funds, labor and physical capital including depreciation). This cost function produces a real marginal lending cost equal to about 5.5%; see Table 6.3 in [161]. This might overestimate the true *short-run* marginal cost since all inputs including labor and physical capital are variable.[17] Recalculating the marginal cost holding real capital and labor constant produces a marginal cost of 4% as is shown in Eq. B.21; we have not had the possibility to reestimate the regression equations.

In Fig. 6.7 we provide a simplified illustration assuming that there is only one bank. Using the Lerner index the uniform lending rate is estimated to 7% assuming uniform monopoly pricing.[18] As above, the short run marginal borrowing cost is estimated to 4%. A small project causes a marginal shift of the demand curve to the right. Even though the project must pay at least 7% to get funding, the marginal cost of providing the funds is 4%, all other things equal. Therefore this analysis suggests that for a marginal project we should discount resources at 4%.[19] On the other hand, a larger project will force up the borrowing and lending rates

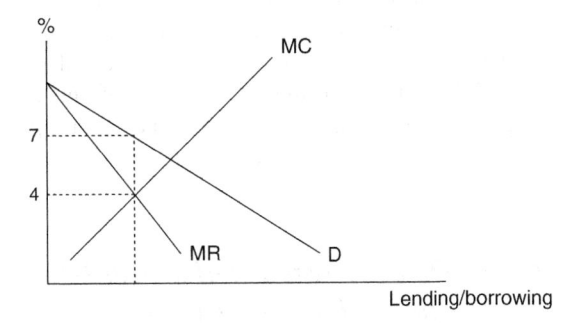

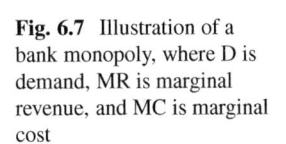

Fig. 6.7 Illustration of a bank monopoly, where D is demand, MR is marginal revenue, and MC is marginal cost

---

[17]For example, taking a Cobb-Douglas cost function with two inputs, the short run MC intersects the long run MC from below at the level of production for which the fixed input is cost minimizing; this holds regardless of whether increasing, constant, or decreasing returns to scale is assumed. Thus if there is slack the short run MC is below its long run counterpart.

[18]The Lerner index is: $(r^L - 5.5)/r^L = 0.22$, where r^L is the lending rate and 5.5 is the (long run) MC estimated by [161]. Thus $r^L \approx 7$.

[19]Strictly speaking, this holds only if the marginal cost curve is more or less horizontal in a small vicinity of the initial optimum.

significantly and hence crowd out consumption (stimulate savings) as well as other projects/investments. The part of the resources that are generated through new savings should roughly speaking be valued at 4+% while crowded out investments should be valued at 7+%.

Productivity has probably increased since the early years of this century, the input price of funds has fallen, and the banking industry is subject to much more intense international competition than 10 years ago. These facts means that the real borrowing cost probably has fallen in comparison to the average for the period covered by [161], ignoring here the likely short-run increase caused by the current economic crisis. Adding the consequences of arbitrage and other possible distortions in financial markets we might end up somewhere around 3% for a small or marginal project as in the point estimate cost-benefit analyses. Thus our point is that firms' high discount rates partly reflect market imperfections in financial markets. However due to the uncertainty about the "correct" discount rate it is important to illustrate the sensitivity of the results to the choice of discount rate, as we also do in Table 6.3.

What is needed to fully resolve these issues is an intertemporal model with different agents having different attitudes to risk and with many different financial instruments and agents and where some agents have market power. Such a model could be used to derive more general evaluation rules than those presented in this book. However, to the best of our knowledge there are no such general models available so their properties are largely unknown. Instead there are different simplified approaches to estimate the social opportunity cost of capital.[20] One typical approach is to define the social opportunity cost of capital as a weighted average of the pre-tax and after-tax rates of return, where the weights reflect the proportions in which financing of a project comes from foregone investment and foregone consumption when there is no market power present. For a recent survey as well as extension the reader is referred to [27] and [28] who also examines the social rate of time preference. The estimates of the social discount rate presented in Sect. 4.3 all draw on different social rate of time preference approaches, since many European authors seem to consider the social opportunity cost of capital as much more volatile over the business cycle.

6.3.2 On the Possibility of a Deal Between Local Residents and the Hydropower Firm

Still, if two parties, in the considered context Fortum and those living in Bollnäs or perhaps the central government, would like to strike a deal it is essentially their business. There is nothing unlawful with such a deal. So if Fortum is willing to sell

---

[20]Under very restrictive assumptions the social opportunity cost of capital and the social rate of time preference coincide, see [27] for details. In such rare cases either concept can be used to define the social discount rate.

3.7 GWh per year forever (our SCENARIO 1) to the municipality of Bollnäs for SEK x million[21] and the municipality is willing to pay this amount a deal is possible. Our pilot study suggests that from the point of view of those currently living in Bollnäs x should not exceed SEK 18 m, but including future generations the point estimate is in excess of SEK 25 m; we do not address the question how to design a real-world payment vehicle such that this amount is collected while respecting individual preferences. If we use the lower bound for the price trajectory and set the discount rate to 6% x would be around SEK 20 m for Fortum, assuming that the firm is risk neutral. Given these assumptions a deal seems possible, at least with respect to SCENARIO 1.

However, still there is an opportunity cost that cannot be escaped. In other words, even if "we", i.e the government or the local community or some other domestic part, have purchased the water, one opportunity is to use the water for electricity generation.[22] Thus there is an opportunity cost of diverting the water from electricity generation to other uses; and we need price forecasts and a discount rate to estimate the expected present value cost less the value of any resources released at Dönje. Moreover, from a societal point of view we must account for any environmental and health impacts of replacement electricity. Then we are back to to the results presented in Sect. 6.1. In the case of a demand reduction we would have to account for the associated present value loss of tax revenue as discussed in Sect. 6.2.3. In other words, the question who pays and how much is more to be seen as a distributional issue separate from the project's real social benefits and costs.

The following should also be noted. It might seem reasonable to believe that a positive sum of willingnesses to pay for a project enables the gainers to hypothetically compensate losers and still having something left over for themselves. Such a project would then imply a hypothetical Pareto improvement. However, [19] showed that even a pure redistribution in a perfect market economy is typically associated with a positive sum of willingnesses to pay (or compensating variations). This result is known as the Boadway paradox and is analyzed in detail by [17]. Thus there are strong arguments against the use of of the sum of unweighted compensating variations in cost-benefit analysis, at least if the considered project is non-marginal.[23] However, for a truly marginal project this particular Boadway paradox vanishes since any induced price effects net out from a general equilibrium cost-benefit rule, as is seen from the rule derived in Appendix A of this book.

---

[21]This amount could be interpreted as a certainty equivalent. That is, the firm is indifferent between receiving this amount with certainty and receiving the extra expected present value profits of continued "full" production at Dönje; see the discussion related to Fig. 6.6.

[22]This assumes that Fortum is willing to sell the electricity within their revenue optimization program given that "we" cover their marginal costs.

[23]One way to state the problem is as follows. Assume that we have a general equilibrium price vector p^* and equilibrium incomes m_h^*, where a subscript h refers to individual h, with the project. Suppose that individual h is willing to pay WTP_h for the project. The vector p^* is not an equilibrium price vector for incomes $m_h^* - WTP_h$, in general. This problem vanishes for the infinitesimally small project which is evaluated at initial or pre-project prices.

Chapter 7
Concluding Remarks

The purpose of this book has been to develop a framework for the societal evaluation of small projects that involve changes in hydropower generation. Such changes cause environmental and other consequences that must be accounted for. By a small or marginal project we mean a project that has virtually no impact on the economy's prices, wages, and exchange rates. There is no exact definition of a small or marginal project but we plan to return to the issue in later studies. For example, if there is an unanticipated emergency stop at a nuclear power plant, one might examine econometrically how the stop affects the spot price and for how long.

Appendix A of this book is devoted to a formal derivation of a set of cost-benefit rules to be used to asses small projects of the kind considered in this study. The rules account for the general equilibrium adjustments of a small open economy. The set of rules is general in the sense that it can be applied to many different kinds of small electricity projects.

This set of rules integrates several key issues, including, but not limited to:

- A contract between the hydropower plant and another party (local residents) generating the general equilibrium cost-benefit rule
- The same contract is a corner stone of our referendum-style Contingent Valuation study
- The tax system in the status quo
- (Partial) foreign ownership of the plant
- Trade in electricity
- Trade in energy certificates
- Trade in carbon emission permits
- Externalities of replacement power (generated in other countries)
- Value of loss of regulating (balancing power) and other system services
- Transmission of electricity modeled as provided by natural monopolies
- Downstream hydrological externalities
- Environmental benefits (aesthetic and otherwise).

In the empirical part of this book we have applied the set of cost-benefit rules to two distinct scenarios. SCENARIO 1 involves a diversion of water during the winter season from electricity generation in order to improve the recreational and other values of the basin downstream the Dönje power plant. In SCENARIO 2 water is also diverted during the summer season. We illustrate numerically most of the different items that show up in a social cost-benefit analysis of the considered changes or projects. However, at this stage we can only provide illustrations since we lack detailed data. Still, we do think that these numerical illustrations serve a purpose. In our opinion they do show how one can proceed in order to obtain more precise estimates. We hope to be able to update with more exact estimates as they become available.

There are some important issues that are not fully addressed in this book. One such issue relates to the treatment of risk and uncertainty. Basically we cover risk through our sensitivity analysis. This is a very applied and rough way of handling risk and could be viewed as implicitly drawing on an assumption that agents are risk-neutral expected utility maximizers. Thus any consequences of risk aversion are omitted except for a brief discussion in Sect. 6.2.5 where risk aversion is addressed. However, it is extremely difficult to say anything definite about the direction and magnitude of the net effect of risk aversion in a multi-agent general equilibrium context. In any case, the issue is left for future research.

Another issue relates to distributional concerns. The approach employed in this study implies that all agents are attributed the same weight. In other words, we simply sum unweighed gains and losses across affected agents. This approach causes a bias in the results unless the welfare distribution in society is optimal. It is far from self-evident that the welfare distribution is optimal. Still, it is not obvious what distributional effects our hypothetical projects would cause if they were implemented. For example, the outcome depends critically on who would compensate the owners of the Dönje plant for their loss of profits. Would it be the local residents living in Bollnäs or would the bill be handed over to the taxpayer at the national level? Recall that the Contingent Valuation study uses a referendum question in order to assess the environmental and other benefits of the scenarios under consideration. It is not necessarily true that a real world implementation of one or the other scenario would be financed by local residents in the way proposed in the Contingent Valuation study. If the project is financed by the government, the distributional outcome depends on what tax, for example a progressive income tax or a capital income tax, that is used to collect the money. In any case, other projects involving changes in water use might produce more obvious groups of winners and losers than the ones under consideration, a fact that motivates that the issue of how to handle distributional issues is addressed in future research.

Appendix A
The Model

All's well that ends well

In this appendix we derive the basic cost-benefit rule employed in this study.[1] We focus on the classic Robinson Crusoe economy where all production and consumption is carried out by a single representative individual. Throughout, solutions to the profit and utility maximization problems, when they exist, are assumed to be unique for each strictly positive price vector. The resulting demand and supply functions are assumed to be continuous and continuously differentiable and all derivatives considered are assumed to be defined and finite. Moreover, we assume there exists a price vector such that all markets simultaneously are in equilibrium (for each fixed level of production at the considered hydropower plant). The reader is referred to, for example, [97] or [135] for discussion of the conditions under which a Walrasian equilibrium with production exists. The static nature of our model means that some dynamic aspects are ignored. In particular, since water can be stored in dams it has a value, referred to as water value. The associated shadow price is determined if the relevant intertemporal maximization problem is solved. The reader is referred to, for example, [58] for a detailed treatment.

In developing the model we assume that there is a single domestically owned hydroelectricity plant (in contrast to the plant at Dönje which is owned by a multinational) which acts as a price taker; multiplying by a number $N > 1$ would not alter the properties of the model but complicate notation. In order to stimulate production of electricity from renewable resources Sweden has introduced what is known as energy certificates. Such certificates are awarded producers of electricity using (some but not all kinds of) renewable resources; they receive one certificate

---

[1] Recently behavioral economics has challenged the assumptions underlying the approach used in this book. For a recent discussion whether behavioral economics has a role for cost-benefit analysis the reader is referred to [165].

P.-O. Johansson and B. Kriström, *The Economics of Evaluating Water Projects*, DOI 10.1007/978-3-642-27670-5, © Springer-Verlag Berlin Heidelberg 2012

for every MWh of "green" electricity they produce. Sellers of electricity to final users must buy energy certificates corresponding to (currently) around 15% of their sales volume.[2] Here it is assumed that the hydropower plant is eligible to receive energy certificates. In order to keep the model as simple as possible we assume that the firm uses only homogeneous labor as an input. In the present case where we are interested in deriving a cost-benefit rule, this simplification involves no loss of generality. The firm aims at maximizing after tax profits,

$$\pi^{hy} = (1 - t^\pi)[(p^s + p^b + p^c) f(\ell^{hy}) - w(1 + t^w)\ell^{hy}], \qquad (A.1)$$

where t^π is a profits tax, p^s is the spot price of electricity, p^b is the extra bonus paid for regulating services provided by the firm (normalized so that it corresponds to the amount received per unit of output x^{hy}), p^c is the price of an energy certificate, $f(.)$ is a production function, ℓ^{hy} is demand for labor, w is the wage rate, and t^w is a social security fee.

The profit function of the firm is defined as follows,

$$\pi^{hy}(.) = (1 - t^\pi)[(p^s + p^b + p^c)x^{hy}(p^s + p^c, w(1 + t^w)) \qquad (A.2)$$
$$-w(1 + t^w)\ell^{hy}(p^s + p^c, w(1 + t^w))],$$

where $x^{hy}(.)$ and $\ell^{hy}(.)$ are functions.

Since the hydroelectricity plant at Dönje is owned by a multinational corporation (Fortum) we consider an open economy. For simplicity there is a single price taking firm in the export sector. We assume this exporting firm faces a maximization problem similar to the one in Eq. A.1. Its profit function expressed in domestic currency is defined as follows,

$$\pi^e(.) = (1 - t^\pi)[p^e e x^e(p^e e, w(1 + t^w)) \qquad (A.3)$$
$$-w(1 + t^w)\ell^e(p^s + p^c, w(1 + t^w))],$$

where a superscript e refers to the export sector, p^e is the export price in foreign currency, e denotes the exchange rate, and $x^e(.)$ and $\ell^e(.)$ are functions.

For the moment, we set aside all other firms of the economy including the Dönje plant. Instead we turn to the maximization problem of the economy's single individual, an assumption which implies we are ignoring distributional issues. The individual values his/her consumption of goods and leisure time. The individual is also affected by emissions from electricity generation. For example, if we produce less electricity at Dönje, we might have to import more electricity generated from coal-fired plants. In addition, the individual derives some benefits from the river downstream the considered power plant, possibly due to activities such as fishing, or canoeing or skating.

---

[2]For a brief description of the Swedish system of energy certificates, see [170].

The Lagrangian of the individual's decision problem is as follows,

$$
\begin{aligned}
\Psi = {} & U(x^d, \mathbf{x}, \ell^s, z^{cli}, z^{rb}, z^b) + \lambda(w(1 - t^\ell)\ell^s + \pi^{hy} + \pi^{tr} \\
& + \pi^e + T - \pi^D + \text{ß} - px^d + \kappa(\pi^F + \pi^D) - \mathbf{p} \cdot \mathbf{x}'),
\end{aligned}
\tag{A.4}
$$

where $U(.)$ is a utility function, x^d is demand for electricity, $\mathbf{x}$ is a vector of other goods, ℓ^s denotes labor supply, z^{cli} represents the welfare impact of emissions, z^{rb} represents the impact on human welfare of the initial flow of water in the considered river basin (through, for example, fish catch, visibility, and other services), z^b is the impact on humans of a particular path for water use, λ is a Lagrange multiplier, w is the wage rate, t^ℓ is a proportional tax on labor income, π^{hy} denotes after-tax profits of a domestic hydroelectricity plant, π^{tr} denotes after-tax profits in transmission of electricity (including any balance services), π^e denotes profits of the economy's single exporter, T is a lump-sum transfer,[3] π^D denotes a transfer to the owner Fortum, of the Dönje power plant, initially equal to zero, but allowed to change so as to induce the owner to reduce production at the Dönje plant, κ is the proportion of Fortum owned by Swedes, π^F denotes initial profits of Fortum, π denotes after-tax profits of other domestically owned firms in the economy, p is the consumer price of electricity, $\mathbf{p}$ is a vector of other prices with the price of a the numéraire commodity normalized to unity, and a prime refers to a transposed vector.

The parameter z^b in the utility function relates to the fact that optimal operation of a hydropower plant implies sharp shifts in water flows. The plant wants to produce at maximum capacity during hours with high prices and then reduce production as prices fall. Similarly, hydropower plants are used to counteract sudden changes in demand or supply. The resulting shifts in water flow cause damage to the river basin and create at times a "moon-like" downstream landscape. The z^b-argument is meant to cover the extent of such damages, i.e. we expect that $dz^b \geq 0$. Similarly, diverting water from electricity generation to the dryway will have a positive impact on fish populations and make the river more attractive for various uses such as canoeing and kayaking as well as provide more aesthetical values. Such values are captured by the z^{rb} variable and we expect that $dz^{rb} > 0$. Finally, any impact on emissions is captured by the z^{cli} variable; if replacement power causes an increase in emissions, $dz^{cli} < 0$ signaling a negative impact on human health and various species that people might care for. Following Eq. A.17 we briefly discuss how to interpret the z-variables in terms of altruistic concerns as well as existence values.

Solving the decision problem in Eq. A.4 yields the indirect utility function, which in a single individual society also is the indirect *social welfare function*,

$$
\begin{aligned}
V = {} & V(p, \mathbf{p}, w(1 - t^\ell), \pi^{hy} + \pi^{tr} \\
& + \pi^e + T - \pi^D + \kappa(\pi^F + \pi^D) + \pi, z^{cli}, z^{rb}, z^b).
\end{aligned}
\tag{A.5}
$$

---

[3] Here all tax rates except T are kept constant; see, however, the discussion following Eq. A.19.

The consumer price of electricity in Eq. A.5 consists of several components. It is defined as follows,

$$p = (p^s + \alpha p^c + p^{tr} + t^{el})(1 + t^{vat}),   \tag{A.6}$$

where α is the proportion of demand for which energy certificates is required (currently around 15%), p^{tr} is the price paid for transmission and regulating services with any fixed component suppressed here, t^{el} is a unit tax on electricity, and t^{vat} is the value added tax.[4]

The government's budget constraint is defined as follows,

$$T = t^{\pi}(\pi^{hy} + \pi^{tr} + \pi^e + \pi^F) + (t^{el} + t^{Vat} p)x^d   \tag{A.7}$$
$$+ t^{Vat} \cdot \mathbf{p} \cdot \mathbf{x}' + (t^{\ell} + t^w)w\ell^s,$$

where t^{Vat} is the value added tax expressed as a proportion of the final consumer price, x^d denotes demand for electricity, and ℓ^s is the supply of labor.

In order to illustrate why sectors other than the electricity-oriented ones will be ignored in what follows, let us consider the partial derivative of V with respect to the producer price p_i of good i. First, by the envelope theorem, see, for example, [97], we have the usual consumer surplus loss $-\lambda x_i^d (1 + t^{vat})$, i.e. the negative of demand for the good times one plus the VAT, converted to welfare units by multiplication by the marginal utility of lump-sum income, here denoted λ. Second, by Hotelling's lemma, profits or producer surplus of the (for simplicity single) firm producing the good increases by x_i^s. Since the single individual of the considered economy owns the firm, his or hers welfare increases by λx_i^s due to this effect. Thus if the price clears the market these two effects will sum to zero if we move the term $-\lambda x_i^d t^{vat}$ to the term reflecting the government's budget (where it will net out against an identical term with opposite sign as can be seen from Eq. A.7). Finally, if demand for the good adjusts, government's VAT revenue will change. Therefore the net impact is equal to,[5]

$$\frac{\partial V}{\partial p_i} = -\lambda \left[(x_i^d - x_i^s) + t^{vat} p_i \frac{\partial x_i^d}{\partial p_i} \right] = -\lambda \left[t^{vat} p_i \frac{\partial x_i^d}{\partial p_i} \right].   \tag{A.8}$$

Since we consider a small change in the level of production at a small plant it will most likely have small secondary effects on demand for other goods. That is, in the CBA it seems reasonable to assume that,

$$\frac{\partial x_i^d}{\partial p_i} \frac{\partial p_i}{\partial \pi^D} \approx 0,   \tag{A.9}$$

---

[4]Throughout it is assumed that the transmission price p^{tr} remains unaffected by the small output changes under consideration.

[5]The same market-clearing result is obtained if we consider $\partial V / \partial p^c$, where p^c is the price of energy certificates. That is, we obtain $-\lambda(\alpha x^d - x^{hy})$ plus the effect on the government's budget. Thus, the market for energy certificates can be ignored as well in the present context.

where $d\pi^D$ is the policy parameter used to generate a change. It is interpreted as a payment to the owner of the Dönje plant such that the owner agrees to marginally reduce the level of production at Dönje. Equation A.9 provides motivation for ignoring induced price effects of the water flow change under consideration outside the electricity market and the export sector. The fact that the export sector cannot be ignored is seen by taking the partial derivative of V with respect to the exchange rate e. One obtains,

$$\frac{\partial V}{\partial e} = \lambda p^e x^e + \frac{\partial V}{\partial T} \frac{\partial T}{\partial e}, \tag{A.10}$$

where, for simplicity, it is assumed that the commodity is not consumed domestically.

The final building block we need is the partial derivative of V with respect to the spot price p^s of electricity,

$$\frac{\partial V}{\partial p^s} = -\lambda(x^d - x^{hy}) + \frac{\partial V}{\partial T} \frac{\partial T^v}{\partial p^s}, \tag{A.11}$$

where the VAT term $x^d t^{vat}$ nets out against an identical term in $\partial T / \partial p^s$, a fact the term T^v in Eq. A.11 accounts for. The difference between demand and domestic supply in Eq. A.11 is equal to import of electricity plus what is produced at the Dönje plant ignoring all other electricity producing plants.

Before turning to the derivation of the cost-benefit rule we will define the parameter $d\pi^D$ ($= (\partial \pi^D / \partial \pi^D) d\pi^D$) as an amount such that,

$$d\pi^D \geq -(1 - t^\pi)[(p^s + p^b)dx^F - w(1 + t^w)d\ell^F + \tag{A.12}$$
$$+ (x^F dp^s - (1 + t^w)\ell^F dw)],$$

where, x^F denotes output at Dönje, ℓ^F denotes labor input, $dp^s = (\partial p^s / \partial \pi^D) d\pi^D$, $dw = (\partial w / \partial \pi^D) d\pi^D$ and, for simplicity, p^b is treated as constant. Thus $d\pi^D$ at least covers the loss of after-tax profits at constant prices plus the after-tax capital gain or loss as production is slightly reduced at Dönje. In what follows, we will assume that Eq. A.12 holds as an equality.[6] This means that we assume that the production level at the Dönje plant as well as all equilibrium prices are continuous and "smooth" functions of the payment scheme π^D.

Equipped with these tools we are now ready to consider the effects of a small change in the parameter π^D,

$$\frac{\partial V}{\partial \pi^D} d\pi^D = \lambda[-d\pi^D - (x^d - x^{hy})dp^s + p^e x^e de \tag{A.13}$$
$$+ (\ell^s - \ell^e - \ell^{hy})dw + dT^1],$$

---

[6]This assumption means that $\kappa(d\pi^F + d\pi^D) = 0$ since Fortum is fully compensated for its loss if production is reduced. This fact is used in deriving the cost-benefit rule below.

where we have moved VAT on electricity consumption and social security fees to the lump-sum term, i.e.,

$$dT^1 = dT - x^d t^{vat} dp^s - (\ell^e + \ell^{hy}) t^w dw,$$

and induced changes in the "environmental" z-variables are suppressed.

Next, let us use Eq. A.12 in Eq. A.13 to obtain,

$$\frac{\partial V}{\partial \pi^D} d\pi^D = \lambda[(p^s + p^b)dx^F - w(1 + t^w)d\ell^F \qquad (A.14)$$
$$-(x^d - x^{hy} - x^F)dp^s + p^e x^e de$$
$$+(\ell^s - \ell^e - \ell^{hy} - \ell^F)dw + dT^2],$$

where

$$dT^2 = dT^1 - t^\pi d\pi^F_{pt} - \ell^F t^w dw,$$

with $d\pi^F_{pt}$ equal to the change of Dönje's pre-tax profits (compare Eq. A.12). If the labor market is in equilibrium, the last but final expression in Eq. A.14 vanishes. Moreover, if the electricity market is in equilibrium then $x^d - x^{hy} - x^F = x^m$, where x^m denotes imports of electricity. Thus Eq. A.14 reduces to,

$$\frac{\partial V}{\partial \pi^D} d\pi^D = \lambda[(p^s + p^b)dx^F - w(1 + t^w)d\ell^F \qquad (A.15)$$
$$+ p^e \cdot x^e de - x^m dp^s + dT^2].$$

Now, if $dp^s = \bar{p}de$, i.e. the international spot price is treated as constant while the exchange rate is flexible, Eq. A.15 contains the initial trade balance in domestic currency $(p^e \cdot x^e - \bar{p}x^m)$ times the change de in the exchange rate. Since the project under consideration is very small, as a rough approximation it seems reasonable to set $de \approx 0$. As a consequence the trade balance change expression vanishes from Eq. A.15. Then after multiplication by $1/\lambda$ Eq. A.15 reduces to,

$$\frac{1}{\lambda} \frac{\partial V}{\partial \pi^D} d\pi^D = (p^s + p^b)dx^F - w(1 + t^w)d\ell^F + dT^2. \qquad (A.16)$$

This equation says that we should evaluate the loss of profits before profits taxes at constant prices. In an open economy context with a firm partly owned by foreigners who "take home" profits it seems as if our approach provides only an approximation; see the discussion following Eq. A.15. However, in cases such as the one under consideration with a very small shift in production, adjustments in exports and imports due to exchange rate adjustments are likely virtually zero. In any case, they are ignored here. A more comprehensive and dynamic model including both stocks and flows is probably needed in order to handle this issue. This is left for future research.

Finally, we must recognize that the "environmental" parameters z^{cli}, z^{rb} and z^b in the indirect utility function also are functions of prices and hence might adjust as we change the parameter π^D. Thus we arrive at a cost-benefit rule,

$$\frac{1}{\lambda}\frac{\partial V}{\partial \pi^D}d\pi^D = -d\pi_c^F + dT^2 + \frac{1}{\lambda}(\frac{\partial V}{\partial z^{cli}}dz^{cli} + \frac{\partial V}{\partial z^{rb}}dz^{rb} + \frac{\partial V}{\partial z^b}dz^b), \quad (A.17)$$

where,

$$d\pi_c^F = (p^s + p^b)dx^F - w(1 + t^w)d\ell^F,$$

so $d\pi_c^F$ is the loss of pre-tax profits for Fortum's Dönje plant evaluated at initial prices. Note that $(1/\lambda)(\partial V/\partial z^i)$, where $i = cli, rb, b$, in Eq. A.17 yields the marginal WTP for a reduction in emissions (but observe that by assumption $dz^{cli} \leq 0$), improved recreational parameters, and a smoother downstream flow of water, respectively. It should be emphasized that the cost-benefit expression, the difference between monetary benefits and monetary costs, is *proportional* to the unobservable change in utility caused by the considered project. Recall that the utility change in Eq. A.17 is converted to monetary units by multiplication by the constant $1/\lambda$.

Any existence values and altruistic concerns are assumed to be captured by the z-variables. To illustrate, if the representative individual is a pure altruist, i.e. respects the preferences of others, $z^{cli} = [z_h^{cli}, \mathbf{U}_{\neq h}(.)]$, where z_h^{cli} denotes any impact of emissions on the representative individual, and $\mathbf{U}_{\neq h}(.)$ is a vector of utility functions of other individuals, possibly including foreigners. Alternatively, the representative individual is a paternalistic altruist, for example, in the sense that he or she is concerned with the health status of other individuals, possibly including non-Swedes. Then z^{cli} will reflect such considerations. If he or she is a pure egoist, $z^{cli} = z_h^{cli}$ since the individual only cares of the impact on his own health. In addition the z^{cli}-variable could be augmented with an argument reflecting existence values if the individual care for endangered species (domestically or globally). The other two z-variables can be given similar interpretations as the z^{cli}-variable. For further details the reader is referred to Sects. 3.2 and 5.1.

The tax term dT^2 in our general equilibrium cost-benefit rule parallels the one found in Eq. 15 in [64]. From Eq. A.7 it can be seen that it has two principal components[7] after the elimination of changes in profits taxes. The first one relates to a change in demand for electricity,

$$[t^{el} + (p^s + \alpha p^c + p^{tr} + t^{el})t^{vat}]\frac{\partial x^d}{\partial p^s}\frac{\partial p^s}{\partial \pi^D}. \quad (A.18)$$

Thus if demand changes (decreases) we should account for tax wedges. We add the unit tax on electricity plus the VAT on the different "parts" of the final electricity price, the spot price plus energy certificates plus transmission price plus unit tax on

---

[7]However, any income effects on demands and supplies in Eq. A.7 of changes in lump-sum items are ignored.

electricity. Note that we do not add the price αp^c paid by the consumer for energy certificates, nor the variable cost p^{tr} for transmission. The reason is the fact that we assume marginal cost pricing of these services. We ignore the possibility that other price changes than the spot price of electricity might affect demand for electricity. If demand falls by the same amount as production at Dönje, we should value the production loss at a "partial" consumer price[8] rather than at producer price. This is seen by letting $dx^d = dx^F$ in Eq. A.18 and using it in Eq. A.16.

Consider next the second principal component of dT^2,

$$(t^\ell + t^w)w\frac{\partial \ell^s}{\partial w}\frac{\partial w}{\partial \pi^D}, \tag{A.19}$$

where we ignore the possibility that other price changes might affect labor supply (any labor needed in transmission, for example). Thus if labor supply decreases by the number of employees released at Dönje, the social marginal cost is the marginal disutility of work effort, $w(1 - t^\ell)$, rather than the gross wage cost $w(1 + t^w)$ to the employer. This is seen by using Eq. A.19 in Eq. A.16 with $d\ell^s = d\ell^F$.

In this book we assume that all tax rates remain constant or unaffected by the considered small projects, so there is a lump-sum tax that adjusts so as to "balance" the government budget. If a tax, say t^{el}, is changed in order to finance the project while lump-sum taxes are ruled out we have a kind of Ramsey problem, see [152]. In this case we have to add the marginal deadweight loss caused by the marginal tax change.[9] For example, with respect to the VAT-component of public revenue in Eq. A.7 one has to add $t^{Vat} \cdot \mathbf{p} \cdot (\partial \mathbf{x}'(.)/\partial t^{el})dt^{el}$ to the cost-benefit rule. The same holds true for any changes in tax revenue from electricity demand[10] and any changes in taxes on labor and profits caused by the adjustment of t^{el} (assuming that at least one good in this economy is untaxed). For more on the Ramsey problem the reader is referred to [45, 69, 140] or [33].

It should be stressed that even a lump-sum tax causes marginal deadweight losses in the presence of distortionary taxation, as should be obvious from the cost-benefit rules derived above. As demand for a taxed commodity adjusts following a change in T there is a change in the associated deadweight loss. Similarly, the parameter $d\pi^D$ affects the deadweight losses (and dT is endogenous in the sense that it is an adjustment caused by our project through $d\pi^D$).

Two related concepts deserve to be briefly discussed. The marginal cost of public funds (MCPF) measures the welfare cost of raising an additional euro in the presence of distortionary taxation. The marginal excess burden (MEB) of taxes is another kind of experiment where, for example, a hypothetical lump-sum

---

[8]We speak of a partial consumer price since, for reasons just described, we exclude $\alpha p^c + p^{tr}$.

[9]Strictly speaking, we here briefly address the more general problem in which there is also the possibility of lump-sum taxation in addition to commodity taxation.

[10]This effect is equal to $(t^{el} + t^{Vat}p)(\partial x^d(.)/\partial t^{el})dt^{el}$ since the direct effect nets out from the cost-benefit rule.

compensation is introduced. This compensation keeps the individual on the same utility level as before a proposed increase in the income tax. According to [6] one can speak of a Harberger-Pigou-Browning tradition or a MEB-tradition in which the marginal cost of public funds is always larger than unity and the Dasgupta-Stiglitz-Atkinson-Stern tradition or MCPF-tradition in which it may be larger or lower than one. In [103] we show that the MCPF becomes complicated to apply in a cost-benefit analysis if prices adjust, as is the case in a general equilibrium context. Moreover, it is partial in the sense that it does not capture the impact of the project, i.e. of $d\pi^D$, on marginal deadweight losses. The MEB refers to a hypothetical experiment and does not seem to be useful for cost-benefit analysis of real world projects, except in a special case where it coincides with the MCPF. The reader is referred to [103] for this result and further discussion. A recent and comprehensive text on the concept of the marginal cost of public funds with many empirical applications is [36].

With respect to environmental and recreational consequences it is simply noted that $dz^{cli} = (\partial z^{cli}/\partial \pi^D)d\pi^D$ most likely has a negative sign. This is so since the electricity production "lost" at the Dönje plant most likely is replaced by electricity generated by fossil-fired power plants. Thus the first term within parentheses in Eq. A.17 reflects a willingness to accept compensation rather than a willingness to pay. For further discussion the reader is referred to the main text.

We end this appendix by briefly considering pricing of transmission services. Typically such services can be characterized as natural monopolies with decreasing average costs. In such circumstances, marginal cost pricing means a loss for the single firm supplying the service in question in a particular region. For a good treatment of the so called access problem, the reader is referred to [3]. A general treatment of nonlinear pricing is provided by [186]. In order to illustrate, let us assume that there are H different consumers. Moreover, assume that the firm sets different two-part tariffs.[11] Its profits is defined as follows,

$$\pi^{tr} = \sum_h [p_h^{tr} x_h^{tr} + \Delta_h^{tr} - c x_h^{tr} - F^{tr}], \tag{A.20}$$

where p_h^{tr} is the variable price paid by consumer h ($h = 1, \ldots .H$), x_h^{tr} is the number of kWh supplied to this consumer, $\Delta_h^{tr} \geq 0$ is the fixed part paid by the consumer, $c > 0$ is the variable unit cost attributed to h, and F^{tr} is the fixed cost per consumer (i.e. F/H, where F refers to the total fixed cost). There is a break-even constraint $\pi^{tr} \geq 0$, and we denote the associated non-negative multiplier μ. We assume that all consumers are equipped with quasi-linear utility functions and that all prices but the transmission price remain constant. Setting aside distributional concerns, the simple objective of our society is assumed to be to maximize the sum of consumer surpluses in the transmission market minus fixed fees plus profits subject to the break-even constraint. Thus the objective is to maximize,

---

[11] We skip here the variation where consumers are offered the choice of different two-part tariffs.

$$W^{tr} = \sum_h \left(\int_{p_h^{tr}}^{\infty} x_h^{dtr}(.) dp_h^{tr} - \Delta_h^{tr} + (1+\mu)[p_h^{tr} x_h^{tr} + \Delta_h^{tr} - cx_h^{tr} - F^{tr}] \right),$$

(A.21)

where $x_h^{dtr}(.)$ is the demand function of consumer h. Consider next the partial derivative of W^{tr} with respect to p_h^{tr}. One obtains,

$$\frac{\partial W^{tr}}{\partial p_h^{tr}} = -x_h^{dtr} + x_h^{tr} + \mu x_h^{tr} + (1+\mu)[p_h^{tr} - c]\frac{\partial x_h^{tr}}{\partial p_h^{tr}},$$

(A.22)

where $x_h^{dtr} = x_h^{tr}$ for all h in equilibrium, i.e. when demand is equal to supply. In an optimum the variable part is such that the derivative in Eq. A.22 is equal to zero assuming here that there is no alternative available, i.e. a consumer cannot quit the scheme. Therefore rearranging the equation one obtains the following Ramsey type of rule,

$$\frac{[p_h^{tr} - c]}{p_h^{tr}} = -\frac{\mu}{(1+\mu)}\frac{(x_h^{dtr}/p_h^{tr})}{(\partial x_h^{dtr}/\partial p_h^{tr})}.$$

(A.23)

If the break-even constraint does not bind ($\mu = 0$) the firm should apply marginal cost pricing. However, if the break-even constraint binds so that $\mu > 0$, then there is a positive markup (ignoring for the moment the question how high the fixed part should optimally be). The magnitude of this markup depends on the multiplier μ and the price elasticity of demand. However, differentiating Eq. A.21 with respect to Δ_h^{tr}, one finds that $\mu = 0$ in optimum so that the variable part is equal to marginal cost[12] in Eq. A.23. From the point of view of the firm it would be optimal to set the fixed fee such that it extracts the entire consumer surplus of h. However, at least in Sweden transmission firms are regulated and they are not allowed to introduce tariffs that are deemed unfair by the regulating body. In any case, assuming that we have transmission pricing according to Eq. A.23 provides justification for ignoring transmission in this book.

---

[12]Thus the firm at least breaks even. Since we look at the sum of consumer surpluses and profits we can not say anything more about the magnitude of the fixed part; it simply works like a transfer from one party to another.

Appendix B
Calculating Present Values: Detailed Presentation

This appendix shows how the different present value items in Chap. 6 were computed.

However, let us first show how the loss of electricity in GWh is estimated. In SCENARIO 1 it is:

$$dGWh^{S1} = -10^{-6} \times 0.85 \times 9.81 \times 33 \times 2.75 \times 4920 \approx -3.7, \tag{B.1}$$

and in SCENARIO 2 it is:

$$dGWh^{S2} = -10^{-6} \times 0.85 \times 9.81 \times 33 \times (2.75 \times 4920 + 10 \times 3840) \approx -14.3. \tag{B.2}$$

Next, let us turn to results presented in Chap. 6. The certainty equivalent price is obtained by finding p^{cert} such that the following equality holds.

$$\int_0^{150} p^{cert} e^{-0.03t} \, dt = 0.5 \times 350 \times \left[\int_0^{150} e^{-0.03t} \, dt \right.$$

$$\left. + \int_0^{20} (1 + 0.05t) e^{-0.03t} \, dt + \int_{20}^{150} (1 + 0.05 \times 20) e^{-0.03t} \, dt \right]. \tag{B.3}$$

The change in present value profits is computed according to the following equation.

$$d\pi_{NPV}^F = dGWh \times 10^3 \times \left[350 \times 0.5 \times \left(\int_0^{150} e^{-0.03t} \, dt + \int_0^{20} (1 + 0.05t) e^{-0.03t} \, dt \right. \right.$$

$$\left. \left. + \int_{20}^{150} (1 + 0.05 \times 20) e^{-0.03t} \, dt \right) - 30 \times \int_0^{40} e^{-0.03t} \, dt \right], \tag{B.4}$$

P.-O. Johansson and B. Kriström, *The Economics of Evaluating Water Projects*,
DOI 10.1007/978-3-642-27670-5, © Springer-Verlag Berlin Heidelberg 2012

where $dGWh$ is the loss of electricity generation in GWh (either -3.7 or -14.3), 350 is the initial spot price in SEK per MWh, the upper limit of integration is set to 150 (years), the growth in the spot price is either 0 or 0.05 for 20 years (and with equal probabilities), the interest rate is 0.03, and the marginal cost is SEK 30 per MWh. A growth equal to 0 (0.05) yields the lower (upper) bound for the loss in revenue.

The present value WTP for a scenario is calculated as follows.

$$WTP_{NPV}^{RB} = 13 \times 10^3 \times 301 \times \int_0^5 e^{-0.03t}\, dt \times (1 + (1/80) \times 32.96), \qquad \text{(B.5)}$$

where 13 is the number of thousands of households in the municipality, 301 is the average annual WTP per household in SEK, expected length of life is 80 years so that 1/80 new generations or 162.5 households enter at each point in time, and 32.96 is the present value of a continuous stream of SEK over the assumed time horizon, i.e. 150 years, when the discount rate is 0.03. In a world with uncertain longevity, we interpret the individual WTP-measure as reflecting a gain in expected utility due to an improved downstream river basin.

To illustrate the WTP-measure in terms of how much present value revenue it might generate it is instructive to turn to discrete time (and arbitrarily it is assumed that payments are made at the beginning of periods).

$$WTP_{NPV}^{Rev} = 13 \times 10^3 \times \sum_{t=0}^4 301 \times 1.03^{-t} - 162.5 \times \left(\sum_{y=1}^4 \sum_{t=y}^4 301 \times 1.03^{-t} \right)$$

$$+ 162.5 \times \left(\sum_{y=1}^{150} \left(\sum_{t=0}^4 301 \times 1.03^{-t} \right) \times 1.03^{-y} \right). \qquad \text{(B.6)}$$

The first term on the right-hand side yields the present value payment made by the current generation given that nobody passes away. The second term yields the loss of present value revenue if 162.5 households on average leaves the cohort each year; recall that households are assumed to pay annually for 5 years. The final term yields the present value payments made by those entering the cohort from year 1 and on, assuming a constant population over time and that all entrants survive until having made their final payment. The sum of these three terms amounts to SEK 26m $(18.5 - 0.4 + 7.6 = 25.7)$. In any case, our claim is that the first term in Eq. B.6 reflects the current population's expected valuation of an improved downstream basin. However, the equation illustrates the fact that actual revenue collection hinges on the design of the payment vehicle.

The present value externality cost of a scenario is calculated as follows.

$$WTA_{NPV}^{E} = -dGWh \times 30500 \times \int_0^{20} e^{-0.03t}\, dt, \qquad \text{(B.7)}$$

where 30,500 is the annual externality cost in SEK per GWh.

The change in investment cost is as follows.

$$I^{\tau} = -170 dGWh \times 10^3 \times \int_{40}^{150} e^{-0.03t} dt. \tag{B.8}$$

The value of the demand change is as follows.

$$dT^{x^d}_{NPV} = dx^d \times \int_{0}^{150} [282 + [480 + 282 + 50 + 200] \times 0.25] \times e^{-rt} dt, \tag{B.9}$$

where $dx^d = dGWh \times 10^3$, and the certainty equivalent spot price is SEK 480 per MWh.

The joint probability density function in Sect. 6.2.5 is as follows when the loss of profits has a uniform distribution while the aggregate WTP is normally distributed and the two variables are independent

$$p_{d\pi,WTP} = \left(\frac{1}{b-a}\right) \times \left(\frac{1}{\sigma_{WTP} \times 2 \times \pi^{1/2}}\right) \times e^{-(WTP - \overline{WTP})^2/(2\sigma^2_{WTP})}, \tag{B.10}$$

where $a = -72$ and $b = -40$ are the limits of integration for the uniform distribution, $\sigma^2_{WTP} = 2.5^2$ is the variance of WTP, and $\overline{WTP} = 26$ is the point estimate of the aggregate WTP.

In the case of a normal distribution for the loss of profits, the joint density function once again assuming independence is as follows,

$$p_{d\pi,WTP} = \left(\frac{1}{2 \times \pi \times \sigma_{WTP} \times \sigma_{d\pi}}\right) \times e^{-0.5 \times \left(\frac{(WTP - \overline{WTP})^2}{\sigma^2_{WTP}} + \frac{(d\pi - \overline{d\pi})^2}{\sigma^2_{d\pi}}\right)}, \tag{B.11}$$

where $\sigma^2_{d\pi} = 25$ is the variance of the loss of profit, and $\overline{d\pi} = -56$ is the point estimate of the profits loss.

In order to arrive at the *cost-benefit acceptability curves* illustrated in Fig. 6.2 we have integrated in the following ways:

$$G_U(x) = 1 - \int_{a+\varpi}^{x} \frac{1}{((b-\varpi) - (a-\varpi))} d(d\pi) \tag{B.12}$$

and

$$G_N(x) = 1 - \int_{-\infty}^{x} \left(\frac{1}{\sigma_{d\pi} \times (2 \times \pi)^{1/2}}\right) \times e^{-(d\pi - \overline{d\pi} + \varpi)^2/(2 \times \sigma^2_{d\pi})} d(d\pi), \tag{B.13}$$

where $G_i(x)$ for $i = U, N$ is a survivor function, $a = -72$, $b = -40$, and $\varpi = WTP_{NPV}^{RB} - WTA_{NPV}^{E} = 24$, i.e. the present value WTP for a scenario less the present value cost of the externality caused by increased emissions from replacement electricity generation. $G_i(x)$ yields the probability that a single observation from a uniform distribution falls in the interval $(x, b + \varpi]$ and the probability that the observation falls in the interval (x, ∞) in the case of a normal distribution. The cost-benefit acceptability curve is traced out by letting x run from the lower limit of integration to $b + \varpi$ in the uniform distribution case (Eq. B.12) illustrated in the left panel of Fig. 6.2 and to $+\infty$ in the normal distribution case (Eq. B.13) shown in the right panel of Fig. 6.2.

In the more general case also WTP_{NPV}^{RB} is a random variable. For notational simplicity, denote the loss of profits by x and the WTP by y and assume they are independent continuous variables. Then the joint survivor function, denoted $G_{x+y}(d\pi)$, can be stated as:

$$G_{x+y}(d\pi) = 1 - F_{x+y}(d\pi) = 1 - \int_{-\infty}^{\infty} F_x(d\pi - y) \times f_y(y)dy, \qquad (B.14)$$

where $F_{x+y}(.)$ is the joint probability distribution function, $F_x(d\pi - y) = \int_{-\infty}^{d\pi-y} f_x(x)dx$, and $f_i(.)$ is a density function with $i = x, y$.

In the case of independent normally distributed variables, $G_{x+y}(d\pi)$, shifted to the left by WTA_{NPV}^{E}, is very close to the survivor function in the right panel of Fig. 6.2; it runs slightly below (above) for probabilities exceeding (falling short of) 0.5. This is so because $x + y$ is $N(\bar{x} + \bar{y}, \sigma_x^2 + \sigma_y^2)$ while $x + \bar{y}$ is $N(\bar{x} + \bar{y}, \sigma_x^2)$, where a bar refers to a mean and σ_i^2 to a variance. Thus, the two approaches produces the same point estimate of the considered project's social profitability. However, the more general approach becomes quite complex if many variables are random. Therefore, we use a simulation approach that is more easily implemented.

The Gumbel (minimum) distribution with support $(-\infty, \infty)$ has probability density function, pdf:

$$p(x) = \frac{1}{s} \times e^{(\frac{x-l}{s} - e^{\frac{x-l}{s}})}, \qquad (B.15)$$

where $p(x)$ denotes the pdf, s is a scale parameter, and l is a location parameter.

In the Gumbel case the survivor function is defined as:

$$G(x) = e^{-e^{\frac{x-l}{s}}}. \qquad (B.16)$$

The cost-benefit acceptability curve is traced out by letting x run from minus infinity to plus infinity in Eq. B.16.

Reference [35] shows that the Gumbel distribution has mean $m = l - s \times \gamma$, where

$$\gamma = \lim_{n \to \infty} \left(\sum_{k=1}^{n} \frac{1}{k} - \ln n \right) \qquad (B.17)$$

is the Euler-Mascheroni constant (≈ 0.5772).

A triangular distribution with support $[a, b]$ and mode c has the density function

$$p(x) = \begin{cases} \frac{2 \times (x-a)}{(b-a) \times (c-a)} & \text{for } a \leq x < c \\ \frac{2 \times (b-x)}{(b-a) \times (b-c)} & \text{for } c \leq x < c \\ 0 & \text{elsewhere} \end{cases} \qquad (B.18)$$

where $a \leq c \leq b$.

The trapezoidal probability density function with support $[a, b]$, lower mode c_l, and upper mode c_u is of the form

$$p(x) = \begin{cases} \frac{u \times (x-a)}{(a-c_l)} & \text{for } a \leq x < c_l \\ u & \text{for } c_l \leq x < c_u \\ \frac{u \times (b-x)}{b-c_u} & \text{for } c_u \leq x < b \\ 0 & \text{elsewhere} \end{cases} \qquad (B.19)$$

where $a \leq c_l \leq c_u \leq b$, and u is the normalizing constant (by which $p(x)$ is a probability density function whose integral over the range equals one) defined as $u = 2(b + c_u - c_l - a)^{-1}$.

Using Eq. 30 in [44], with $\alpha = 1$ and $\eta_1 = \eta_3 = 2$, the mean for the trapezoidal distribution can be calculated as follows

$$m = \frac{(c_l - a) \times (a + 2 \times c_l)/3 + (c_u^2 - c_l^2) + (b - c_u) \times (2 \times c_u + b)/3}{(c_l - a) + 2 \times (c_u - c_l) + (b - c_u)}, \qquad (B.20)$$

where m denotes the mean. Setting $c_l = c_u = c$, i.e. assuming a triangular distribution, the reader can verify that Eq. B.20 reduces to $m = (a + b + c)/3$ while if $c_l = a$ and $c_u = b$, i.e. assuming a uniform distribution, Eq. B.20 simplifies to $m = (b - a) \times (b + a)/(2 \times (b - a)) = (a + b)/2$. It should be added that the generalized trapezoidal distribution is available in the free statistical software R.

The short-run marginal cost in Sect. 6.3.1 is estimated as:

$$\frac{dc}{dq} = \frac{e^{ln(c)}}{q} \times [s_0 + s_1 \times ln(q) + s_2 \times ln(\omega_1)] =$$

$$\frac{1276}{24823} \times (0.6134 - 0.00059 \times ln(24823) - 0.04658 \times ln(0.0299)) = 0.0396, \qquad (B.21)$$

where c refers to total cost, q to output (total assets), ω_1 to input cost of funds, s_0, s_1, and s_2 are estimated coefficients; see Tables 5.1, 5.2 and 6.1 in [161]. If labor is considered variable too, dc/dq increases to 0.0417. The study's average input price of funds might be a reasonable lower bound for the short-run marginal cost. If so dc/dq is at least 0.03 for the considered period; see ω_{1i} in Sjöström's Table 5.2.

The remaining items in the sensitivity analysis follows from the above equations.

Appendix C
Spot Prices

Below are listed (vertically) monthly average spot prices (system price) in SEK at the Nord Pool market since January 1996–2010 (inclusive); the first column contains prices for 1996–1999, the second column 2000–2003, the third 2004–2007, and the fourth 2008–2010.

197,89	139,31	265,15	431,94
269,99	109,72	252,39	360,92
241,65	98,81	269,44	278,32
252,62	105,89	263,87	354,98
269,51	78,33	254,53	240,27
259,63	86,70	292,78	379,07
249,24	53,39	258,79	420,05
308,96	82,10	299,95	513,30
337,03	119,18	263,45	644,50
292,22	131,38	251,48	554,87
234,63	144,53	262,76	518,62
244,31	146,90	232,87	478,51
234,39	181,87	208,19	444,41
175,77	242,34	228,15	414,40
139,99	235,78	267,65	392,25
135,37	240,67	281,59	370,68
121,01	217,83	284,03	345,68
117,18	232,43	242,74	384,80
92,23	209,20	271,78	356,89
137,49	198,27	289,63	331,04
107,13	201,66	274,10	291,86
136,79	182,78	302,86	348,04
169,37	201,43	291,50	375,87
185,82	222,78	323,23	412,54
174,60	226,11	375,60	544,37
157,52	186,40	405,03	684,66

P.-O. Johansson and B. Kriström, *The Economics of Evaluating Water Projects*,
DOI 10.1007/978-3-642-27670-5, © Springer-Verlag Berlin Heidelberg 2012

137,64	168,41	492,02	554,72
127,31	158,52	483,11	453,26
111,71	140,83	351,09	414,90
124,36	149,74	406,83	428,37
72,24	144,98	456,29	431,73
57,38	187,36	612,40	404,06
84,38	225,78	589,87	455,83
114,38	285,23	496,50	460,15
153,49	392,12	425,81	510,63
160,46	677,59	302,09	740,02
145,21	656,04	250,13	
130,36	441,67	264,47	
110,55	364,16	221,48	
94,03	288,91	207,32	
101,33	270,15	196,65	
88,54	226,03	220,86	
72,82	254,19	162,13	
116,68	304,85	154,02	
138,95	293,32	234,28	
141,62	316,39	335,54	
132,35	326,16	424,49	
148,82	280,26	433,17	

Appendix D
The Survey Questions

Below is a compact summary of the questionnaire.

1. Are you a member of a environmental organization or similar? Which?
2. Does your household buy certified electricity
3. Summer activities in rivers or lakes. # of occasions

   a. Fishing
   b. Swimming.
   c. Ice-skating
   d. Hiking, running or cycling.
   e. Boating.

4. Winter activities in rivers or lakes. # of occasions

   a. Fishing
   b. Swimming.
   c. Ice-skating
   d. Hiking, running or cycling.
   e. Boating.

5. What type of boat (if any) do you use?

   a. Kayak – Type?
   b. Rowboat-Type?
   c. Canoe-Type?
   d. River-rafting-Boat type?
   e. Sailing boat-type?
   f. Motorboat – type?
   g. Steamboat-type

6. Summer activities in Ljusnan. # of occasions

   a. Fishing
   b. Swimming.
   c. Ice-skating

d. Hiking, running or cycling.
e. Boating.

7. Winter activities in Ljusnan. # of occasions

a. Fishing
b. Swimming.
c. Ice-skating
d. Hiking, running or cycling.
e. Boating.

8. Do you live close to the Bollnäs streams? What distance?
9. Do you own property along the Bollnäs streams?
10. Have you ever driven, biked or walked across the Klumpströmmen bridge (marked on the map).
11. Description of impacts of the scenario

a. Positive impact on ecological system... (details omitted)
b. More fish (species and stock development description)... (details omitted).
c. Landscape attractiveness... (details omitted)

WTP-question introduction

It has become more common in Sweden that those who are affected by local environmental issues are able to vote in local referendums. The following proposal can be viewed as such a local referendum. It is thought to be held among inhabitants of Bollnäs municipality. The purpose of this question is to shed some light on how the average citizen of Bollnäs values a potential change of the water flow in the Bollnäs streams. The change will improve fishing conditions, water ecology and landscape aesthetics. At present, the water rights are owned by the Fortum company. This means that Fortum has the right to produce electricity at the Dönje plant. Suppose that the only possible way to increase the water flow in the Bollnäs streams is to buy back those water rights, by means of a joint action among Bollnäs citizens.

Proposal to change of the winter season water flow in the Bollnäs streams.

*The **Proposal** entails an increase of the winter season water flow from 0.25 m^3s^{-1} to 3 m^3s^{-1}. There will be no change in the summer season water flow. This means that the water flow will increase from the power station to Varpen (note: this was described in a map not included here) The proposal is depicted in a series of pictures in the sequel. The total costs for the **Proposal** is not known with certainty at the present time. Suppose that the referendum is held when the cost has become known. If a majority supports the proposal, it will be undertaken. A "yes" entails each household paying a given sum over a period of 5 years.* A series of figures were then introduced in order to depict SCENARIO 1, see Figs. 5.1 and 5.2.

12. Would you be prepared to pay anything for the proposal? If you are unsure about your exact WTP, you can use an interval of choice.
13. State the largest sum that you would be prepared to pay for the proposal. (Ask about lower and upper if choose interval)
14. I vote "no" to the proposal independently of the cost because (state reasons)
15. Gender
16. Year of birth
17. Postal code
18. Achieved level of education
19. Employment
20. Household income net of tax
21. # of Children < 18 years old in household
22. # of Persons > 18 years old in household
23. # of Persons > 18 years old in household that contribute to household income

Appendix E
Data

	lower	wtp	upper
1	100	3000	500
2	1200	100	2500
3	50	150	100
4	100	100	200
5	100	500	200
6	0	200	5000
7	1	100	25
8	100	200	300
9	500	300	1500
10	100	600	500
11	50	1000	200
12	50	100	200
13	100	300	500
14	100	100	500
15	500	300	1500
16	50	50	200
17	100	1000	200
18	100	1000	300
19	500	500	1500
20	100	2500	500
21	1000	100	1500
22	100	1000	500
23	25	1000	75
24	0	500	1200
25	500	200	1000
26	600	150	1200
27	1	100	300
28	50	200	150
29	0	500	2000
30	100	10	200

P.-O. Johansson and B. Kriström, *The Economics of Evaluating Water Projects*,
DOI 10.1007/978-3-642-27670-5, © Springer-Verlag Berlin Heidelberg 2012

31	0	200	1000
32	100	1	400
33	10	500	15
34	400	400	800
35	0	300	500
36	1	100	50
37	50		100
38	100		200
39	0		1000
40	25		150
41	10		100
42	1000		2000
43	300		700
44	100		500
45	100		200
46	100		1200
47	1500		3000

Appendix F
Programs

The estimations are carried out using R. To download this free program, go to www. r-project.org.

F.1 Reading the Data

```
bol<-read.table("donje_interval_data.txt",header=F) # Data above
bol.ex<-bol[!is.na(bol[,3]),3] # point observations
length(bol.ex) ## 36
bol.inv<-bol[order(bol[,2],bol[,4]),c(2,4)] ##ordered intervals
dim(bol.inv) # 47 2
## Zero observations 135-47-36=52
```

F.2 Jammalamadaka Estimator

```
##Jammalamadaka's non-parameteric estimate.
NPE4IC<-function(obs=rpois(100,100),int=matrix(rpois(100,100)
        ,nc=2,byrow=T),dist=NPE0(obs,int)){
n1<-length(obs)
n2<-nrow(int)
n<-n1+n2
obs<-sort(obs)
int<-int[order(int[,1],int[,2]),]
fstep<-stepfun(dist[,1],c(0,dist[,2])) ## A function
newF<-matrix(nr=nrow(dist),nc=2)
newF[,1]<-dist[,1]
i<-0
for (k in c(dist[,1])){
i<-i+1
term1<- rank(c(k,obs),tie="max")[1]-1 ## OBS: ties possible.
term2<- rank(c(k,int[,2]),tie="max")[1]-1 ## #of intervals to the left
term3<-0
```

```
if (k < max(int[,2])){
ind<- which(k>int[,1] & k< int[,2])
if (length(ind)>0) {
for (j in ind)
term3<-term3+ (fstep(k)-fstep(int[j,1]))/(fstep(int[j,2]-0.000001)
              -fstep(int[j,1]))## epsilon<- 0.000001
}
}
newF[i,2]<- (term1+term2+term3)/n
}
newF
}
save.image("BolOnly.Rdata")
cdf0<-NPE0(bol.ex,bol.inv) ## initial cumulative distribution function (cdf)
## stop if the max difference <0.005 (arbitrary, but fairly small number)
cdf1<-NPE4IC(bol.ex,bol.inv,cdf0); max(abs(cdf1-cdf0)) #0.07624
cdf2<-NPE4IC(bol.ex,bol.inv,cdf1); max(abs(cdf1-cdf2)) #0.0307
cdf3<-NPE4IC(bol.ex,bol.inv,cdf2); max(abs(cdf3-cdf2)) #0.0141
cdf4<-NPE4IC(bol.ex,bol.inv,cdf3); max(abs(cdf3-cdf4)) #0.0065
cdf5<-NPE4IC(bol.ex,bol.inv,cdf4); max(abs(cdf5-cdf4)) #0.0003 Stopped.
```

F.3   Estimating the Weibull Distribution

```
lweibull<-function(par,obs,int){
    n1<-length(obs)t
    n2<-nrow(int)
    n<-n1+n2
    k<-par[1]
    lamda<-par[2]
     logL<-n1*log(k/lamda)+(k-1)*sum(log(obs))-n1*(k-1)*log(lamda)
     -sum((obs/lamda)^k)+sum(log(exp(-int[,1]^k/lamda^k)
     -exp(-int[,2]^k/lamda^k)))
     -logL
}
library(STAR)
pars<-weibullMLE(bol.ex)$estimate
> optim(as.numeric(pars),fn=lweibull,obs=bol.ex,
  int=bol.inv,lower=c(0.001,0.1),upper=c(3,1000))
$par
[1]   0.8613032 389.7161518
# Converges for many other starting values to the same vector
```

F.4   Programs for the Belyaev-Kriström Estimator

```
install.package("iwtp") #Get package
install.packages("intervals") # useful package (optional)
# Take data from end of the book and make a dataframe
intrv.data=read.table("donje_interval_data.txt")
# Keep only the intervals
```

```
intervals.df=data.frame(1:47,intrv.data[,1],intrv.data[3])
# Calculate tn, number of repeated intervals. See below for source
source("intervme.R")
my.data=interv.me(intervals.df)
as.data.frame(my.data)
# bounds on Weibull
bounds <- list(a=c(lower=50,upper=500,init=200),
b=c(lower=0.5,upper=4.5,init=0.8))
# Weibull\ldotskeep likelihood value\ldots bm=1
# refers to behavioral model 1 and so on. See below
m1=iwtp(my.data, dist = "weibull",bm=1,bounds=bounds)
llik1=m1$sf[2]
m2=iwtp(my.data, dist = "weibull",bm=2,bounds=bounds)
llik2=m2$sf[2]
m3=iwtp(my.data, dist = "weibull",bm=3,bounds=bounds)
llik3=m3$sf[2]
m4=iwtp(my.data, dist = "weibull",bm=4,bounds=bounds)
llik4=m4$sf[2]
m5=iwtp(my.data, dist = "weibull",bm=5,bounds=bounds)
llik5=m5$sf[2]
lliks=c(llik1,llik2,llik3,llik4,llik5)
lliks
# plot final model
plot(m5,col="black")
```

F.4.1   Helper Function intervme.R

```
tab2matrix <- function(tab=null){
    mat.intv <- matrix(c(0,0,0), ncol=3)
    for(i in 1:length(tab[,1])) {
        for(j in 1:length(tab[1,])) {
            if(tab[i,j] >0) {
                tmp.row <- c(rownames(tab)[i], colnames(tab)[j], tab[i,j])
                mat.intv <- rbind(mat.intv, as.integer(tmp.row))
            }
        }
    }
#mat.intv
    mat.intv <- mat.intv[-1,]
    mat.intv <- cbind(c(1:length(mat.intv[,1])), mat.intv)
}


##

interv.me <- function(data) {
    # data check
    if (is.null(data))
        stop("argument data must be specified!",call.=FALSE)

    if (!is.data.frame(data)) {
        stop("data must be a dataframe!",call.=FALSE)
    }
```

```
    if(dim(data)[2]<3)
        stop("The input data should have at least 3 columns!",call.=FALSE)

    tab <- table(data[,2],data[,3])
    data.mat <- tab2matrix(tab)
    data.df <- as.data.frame(data.mat)
    names(data.df) <- c("respondent no","lower value", "upper value","th")

    out <- data.df
    return(out)
}
```

F.4.2   Behavioral Model BM1..BM5

These models are automatically included in iwtp, but are printed here for completeness. Programs by Zhou Wenchao, Department of Forest Economics, SLU-Umeå.

```
## define generic function Dofreq
Dofreq <- function(x,\ldots) {
    UseMethod("Dofreq")
}


#
# S3 method
# estimates for frequency based on attractive
# argument x is a object of class 'divs.BM1',
# argument intv is a object of class 'interv',
#an output from function interv
## BM1: is the model of indifferent respondents with whj=1/dj
Dofreq.divs.BM1 <- function(x, intv, limits=list()) {
    # bof local functions
    tiesfordivisions <- function(intv) {
        mu <- length(intv$n.intv)
        kv <- dim(intv$divs)[1]
        ndx <- intv$ndx.divs
        pa <- cbind(intv$n.intv,ndx)
        mat <- t(apply(pa,1,function(xi,kl=0)
        {v <- rep(0,kl);v[xi[2]:xi[3]]<- 1;
        return(v)},kl=kv))
        rownames(mat) <- c(1:mu)
        #print(mat)
        return (mat)
    }
    # eof local functions

    ties <- tiesfordivisions(intv) #
    freqs <- t(t(ties)/colSums(ties))
out <- list(intv=intv,freq.mode="BM1",
coef=list(),coef.limits=list(),freqs.divs=freqs)
    class(out) <- "iwtp.interv"
    return(out)
}
```

```
#
# S3 method
# estimates for frequency based on the last div
# argument x is a object of class 'divs.BM2',
# argument intv is a object of class 'interv', an output
#from function interv
## BM2 is the model of respondents
#who with vj containing their WTP-point
## select to state uh in which vj is the last division interval.
##
Dofreq.divs.BM2 <- function(x, intv,limits=list()) {
    # local functions
    tiesforHiDivision <- function(intv) {
        mu <- length(intv$n.intv)
        kv <- dim(intv$divs)[1]
        ndx <- intv$ndx.divs
        pa <- cbind(intv$n.intv,ndx)
        mat <- t(apply(pa,1,function(xi,kl=0)
  {v <- rep(0,kl);v[xi[3]]<- xi[1];return(v)},kl=kv))
        rownames(mat)  <- c(1:mu)
        #print(mat)
        return (mat)
    }
    # end of local functions

    ties <- tiesforHiDivision(intv)  # select only the last devision

    freqs <- t(t(ties)/colSums(ties))
    freqs[is.na(freqs)] <- 0

    out <- list(intv=intv,freq.mode="BM2",coef=list(),
coef.limits=list(),freqs.divs=freqs)
    class(out) <- "iwtp.interv"
    return(out)
}


#
# S3 method
# argument x is a object of class 'divs',
# argument intv is a object of class 'interv',
#an output from function interv
## BM3: is the model of respondents who,
#with vj containing their wtp point, select
## to state uh, proportionally to achoring
#probabilities wh, h = 1, \ldotsm
Dofreq.divs.BM3 <- function(x, intv, limits=list()) {
    # bof local functions
    tiesfordivisions <- function(intv) {
        mu <- length(intv$n.intv)
        kv <- dim(intv$divs)[1]
        ndx <- intv$ndx.divs
        pa <- cbind(intv$n.intv,ndx)
mat <- t(apply(pa,1,function(xi,kl=0)
  {v <- rep(0,kl);v[xi[2]:xi[3]]<- xi[1];return(v)},kl=kv))
        rownames(mat)  <- c(1:mu)
```

```
                #print(mat)
                return (mat)
    }
    # eof local functions

    ties <- tiesfordivisions(intv) #
    freqs <- t(t(ties)/colSums(ties))
    out <- list(intv=intv,freq.mode="BM3",
coef=list(),coef.limits=list(),freqs.divs=freqs)
    class(out) <- "iwtp.interv"
    return(out)
}


#
# S3 method
# argument x is an object of class 'divs.ranks',
# which is a list containing the #coefficient
# argument intv is an object of class 'interv',
# an output from function interv
# BM4: beta-type weighting of acnchoring probabilities wh, h=1, \ldots, m
##
Dofreq.divs.BM4 <- function(x,intv,
        limits=list(
                c1=c(lower=0.01,upper=100,init=0.1),
                c2=c(lower=0.01,upper=10,init=0.001))) {
    # bof local function
    ranksfordivisions <- function(intv) {
        mu <- length(intv$n.intv)
        kv <- dim(intv$divs)[1]
        ndx <- intv$ndx.divs
        ranks <- t(apply(ndx,1,function(xi,kl=0){v<-rep(0,kl)
;n<-xi[2]-xi[1]+1;v[xi[1]:xi[2]]<-1:n;return(v)},kl=kv))
        rownames(ranks) <- c(1:mu)
        #print(ranks)
        return (ranks)
    }

do.frequencies <- function(th,ranks,coef.ranks=0,size.divs) {
        freq.div <- function(tr,th,size,coef=0) {
                # estimator of w
                w <- th/sum(th)
                # sum rank over division j,
                #sum.tr <- sapply(size, function(x) sum(1:x))
                last.tr <- sapply(size, function(x) x)
                whj <- rep(0,length(th))

# apply only to stated intervals u which contains division j
                idx <- which(tr>0)
                # calculate whj
                if(length(coef)==1)
                    coef = c(coef,1)

                #
                rcj <- rep(0,length(th))
                #rcj[idx] <- tr[idx]/sum.tr[idx]
                rcj[idx] <- tr[idx]/last.tr[idx]
```

```
whj[idx] <- rcj[idx]^(coef[1]-1) * ifelse((1-rcj[idx])==
0,1,(1-rcj[idx])^(coef[2]-1))*w[idx]
            #

            sw <- sum(whj[idx])
            return(whj/sw)
        }

feq <- apply(ranks,2,freq.div,th=th,size=size.divs,coef=coef.ranks)
     return(feq)
    }

    # eof local functions

    ranks <- ranksfordivisions(intv)
    coef <- unlist(x)
    names(coef) <- names(x)
freqs <- do.frequencies(intv$n.intv,ranks,coef.ranks=coef,intv$size.divs)

    out <- list(intv=intv,freq.mode="BM4",coef=coef
,coef.limits=limits,freqs.divs=freqs)

    class(out) <- "iwtp.interv"
    return(out)
}

# S3 method
# argument x is an object of class 'divs.ranks',
#which is a list containing the coefficient
# argument intv is an object of class 'interv',
# an output from function interv
# BM5: proportionally to wh*(1+c(rhj/rh.)))
##
Dofreq.divs.BM5 <- function(x,intv,
        limits=list(
                c=c(lower=0.01,upper=100,init=0.1))) {
    # bof local function
    ranksfordivisions <- function(intv) {
        mu <- length(intv$n.intv)
        kv <- dim(intv$divs)[1]
        ndx <- intv$ndx.divs
        ranks <- t(apply(ndx,1,function(xi,kl=0)
{v<-rep(0,kl);n<-xi[2]-xi[1]+1;v[xi[1]:xi[2]]
<-1:n;return(v)},kl=kv))
        rownames(ranks) <- c(1:mu)
        #print(ranks)
        return (ranks)
    }

do.frequencies <- function(th,ranks,coef.ranks=0,size.divs) {
    freq.div <- function(tr,th,size,coef=0) {
            # estimator of w
            w <- th/sum(th)
            # sum rank over division j,
            sum.tr <- sapply(size, function(x) sum(1:x) )
            whj <- rep(0,length(th))
```

```
# apply only to stated intervals u which contains division j
            idx <- which(tr>0)
            # calculate whj

            # only one c
whj[idx] <- w[idx]*(1+coef[1]*(tr[idx]/sum.tr[idx]))

            sw <- sum(whj[idx])
            return(whj/sw)
        }

feq <- apply(ranks,2,freq.div,th=th,size=size.divs,coef=coef.ranks)
    return(feq)
    }

    # eof local functions


    ranks <- ranksfordivisions(intv)
    coef <- unlist(x)
    names(coef) <- names(x)
freqs <- do.frequencies(intv$n.intv,ranks,coef.ranks=coef,intv$size.divs)

    out <- list(intv=intv,freq.mode="BM5",
coef=coef,coef.limits=limits,freqs.divs=freqs)

    class(out) <- "iwtp.interv"
    return(out)
}
```

Bibliography

1. Adamowicz, W.: What's it worth? an examination of historical trends and future directions in environmental valuation. Aust. J. Agric. Resour. Econ. **48**, 1467–8489 (2004)
2. Amundsen, E.S., Bergman, L.: Why has the nordic electricity market worked so well? Util. Policy **14**, 148–157 (2006)
3. Armstrong, M., Doyle, C., Vickers, J.: The access pricing problem: a synthesis. J. Ind. Econ. **2**, 131–150 (1996)
4. Aronsson, T., Johansson, P.O., Löfgren, K.G.: Welfare Measurement, Sustainability and Green National Accounting. Edward Elgar, Cheltenham (1997)
5. Arrow, K.J., Fisher, A.C.: Environmental preservation, uncertainty, and irreversibility. Q. J. Econ. **88**, 312–319 (1974)
6. Ballard, C.L., Fullerton, D.: Distortionary taxation and the provision of public goods. J. Econ. Perspect. **6**, 117–131 (1992)
7. Barro, R., Sala-I-Martin, X.: Economic Growth, 1st edn. McGraw-Hill, Cambridge (1995)
8. Bateman, I.: Economic Valuation with Stated Preference Techniques: A Manual. Edward Elgar, Cheltenham (2002)
9. Bateman, I.J., Carson, R.T., Day, B., Hanemann, M.W., Hanley, N., Hett, T., Jones-Lee, M., Loomes, G., Mourato, S., Ozdemiroglu, E., Pearce, D.W., Sugden, R., Swanson, J.: Economic Valuation With Stated Preference Techniques. Edward Elgar, Cheltenham (2004)
10. Bell, K.P., Huppert, D., Johnson, R.L.: Willingness to pay for local Coho salmon enhancement in coastal communities. Mar. Resour. Econ. **18**, 15–31 (2003)
11. Belyaev, Y., Kriström, B.: Approach to analysis of self-selected interval data. Revised: CERE Working Paper # 2. http://www.cere.se/documents/wp/CERE_2010_2.pdf (2009)
12. Belyaev, Y., Kriström, B.: Statistical models for analysis of data with self-selected censoring intervals. Tech. Rep., 2010 Annual Meeting of the Institute of Mathematical Statistics (IMS), Göteborg, 9–13 Aug 2010
13. Bennett, J., Blamey, R.: The Choice Modelling Approach to Environmental Valuation. Edward Elgar, Cheltenham (2001)
14. Berrens, R.P., Bohara, A.K., Jenkins-Smith, A.K., Silva, C.L., Ganderton, P., Brookshire, D.: A joint investigation of public support and public values: case of instream flows in New Mexico. Ecol. Econ. **27**(2), 189–203 (1998)
15. Birol, E., Karousakis, K., Koundouri, P.: Using economic valuation techniques to inform water resources management: a survey and critical appraisal of available techniques and an application. Sci. Total Environ. **365**(1–3), 105–122 (2006)
16. Bishop, R.C., Boyle, K., Welsh, M.P., Baumgartner, R.M., Rathbun, P.R.: Glen Canyon dam releases and downstream recreation. Glen Canyon Environmental studies. Tech. Rep. Bureau of Reclamation, Salt Lake City (1987)

17. Blackorby, D., Donaldson, C.: A review article: the case against using the sum of compensating variations in cost-benefit analysis. Can. J. Econ. **23**, 471–494 (1990)
18. Blanchard, O., Fisher, S.: Lectures on Macroeconomics. MIT, Cambridge, MA (1996)
19. Boadway, R.W.: The welfare foundations of cost-benefit analysis. Econ. J. **84**, 926–939 (1974)
20. Boadway, R.W., Bruce, N.: Welfare Economics. Basil Blackwell, Oxford (1984)
21. Boland, J., Flores, N., Howe, C.: The theory and practice of benefit-cost analysis. In: Russell, R., Baumann, D. (eds.) The Evolution of Water Resource Planning and Decision Making, pp. 82–135. Edward Elgar, Cheltenham (2009)
22. Boyle, K., Bishop, R., Caudill, J., Charbonneau, J., Larson, D., Markowski, M., Unsworth, R., Patterson, R.: A meta analysis of sport fishing values. Tech. Rep., U.S. Fish and Wildlife Service, Washington, DC (1999)
23. Brännlund, R., Kriström, B.: Miljöekonomi. Studentlitteratur, Lund (1998)
24. Briggs, A.H., Sculpher, M.J., Claxton, K.: Decision modelling for health economic evaluation. Oxford University Press, Oxford (2006)
25. Broberg, T., Brännlund, R.: An alternative interpretation of multiple bounded wtp data-certainty dependent payment card intervals. Resour. Energy Econ. **30**, 555–567 (2008)
26. Bunn, D.W., Karaktatsani, N.: Forecasting Electricity Prices. Mimeo, London Business School, London (2003)
27. Burgess, D.: Removing some dissonance from the social discount rate debate. Working Paper 2008-2, Department of Economics, University of Western, Ontario, 2008
28. Burgess, D.F., Zerbe, R.O.: Appropriate discounting for benefit-cost analysis. J. Benefit-Cost Anal. **2**(2), Article 2 (2011)
29. Carbone, J.C., Smith, V.K.: Evaluating policy interventions with general equilibrium externalities. J. Public Econ. **92**(5–6), 1254–1274 (2008)
30. Carbone, J.C., Smith, V.K.: Valuing ecosystem services in general equilibrium. WP 15844, National Bureau of Economic Research, Cambridge (2010)
31. Clawson, M.: Methods of measuring the demand for and value of outdoor recreation. Resources for the Future Reprint No. 10, Washington, DC (1959)
32. Clawson, M., Knetsch, J.L.: The values of land and water resources when used for recreation. In: Clawson, M., Knetsch, J.L. (eds.) Economics of Outdoor Recreation. Johns Hopkins University Press, Baltimore (1966)
33. Coady, D., Drèze, J.: Commodity taxation and social welfare: the generalised Ramsey rule. Tech. Rep., London School of Economics, London (2000)
34. Coase, R.H.: The problem of social cost. J. Law Econ. **3**, 1–44 (1960)
35. Coles, S.G.: An Introduction to Statistical Modeling of Extreme Values. Springer, New York (2001)
36. Dahlby, B.: The Marginal Cost of Public Funds. Theory and Applications. MIT, Cambridge, MA (2008)
37. Dasgupta, P., Mäler, K.G.: The environment and emerging development issues. Beijer Reprint Series, No. 1, The Royal Swedish Academy of Sciences, Stockholm (1991)
38. Dasgupta, P., Maskin, E.: Uncertainty and hyperbolic discounting. Am. Econ. Rev. **95**, 1290–1299 (2005)
39. Davis, R.K.: The value of outdoor recreation. An economic study of the main woods. Ph.D. Thesis, Harvard university (1963)
40. de Rus, G.: Introduction to Cost-Benefit Analysis. Looking for Reasonable Shortcuts. Edward Elgar, Cheltenham (2010)
41. Diamant, A., Willey, Z.: Water for salmon: an economic analysis of salmon recovery alternatives in the Lower Snake and Columbia Rivers. Report by the Environmental Defense Fund to the Northwest Power Planning Council, EDF (1995)
42. Dixit, A.K., Pindyck, R.S.: Investment Under Uncertainty. Princeton University Press, Princeton (1994)
43. Dobes, L.: A century of Australian cost-benefit analysis: lessons from the past and the present. Working Paper 2008-01, Office of Best Practice Regulation, Department of Finance and Deregulation (2008)

44. van Dorp, R.J., Kotz, S.: Generalized trapezoidal distributions. Metrika **58**, 85–97 (2003)
45. Drèze, J., Stern, N.: The theory of cost-benefit analysis. In: Auerbach, A., Feldstein, M. (eds.) Handbook of Public Economics. North-Holland, Amsterdam (1987)
46. Duffield, J.W., Brown, T.C., Allen, S.D.: Economic value of instream flow in Montana's big hole and bitterroot rivers. Research Paper RM-317, USDA Forest Service (1994)
47. Eckstein, O.: Water Resource Development. Harvard University Press, Cambridge, MA (1958)
48. Ekström, M.: Nonparametric estimation for classic and interval open-ended data in contingent valuation. Research Report 7, Centre of Biostochastics, Swedish University of Agricultural Sciences, Uppsala (2010)
49. EPA: Guidelines for Preparing Economic Analyses. U.S. Environmental Protection Agency (EPA), Washington, DC (2000). EPA 240-R-00-003
50. European Commission: Guide to cost-benefit analysis of investment projects. Tech. Rep., DG Regional Policy (2008)
51. Evans, D.J.: Social discount rates for the European union. In: Florio, M. (ed.) Cost-Benefit Analysis and Incentives in Evaluation. Edward Elgar, Cheltenham (2007)
52. Feldstein, M.S.: The derivation of social discount rates. Kyklos **18**, 277–287 (1965)
53. FERC: Final supplemental final environmental impact statement, Condit hydroelectric project washington. Tech. Rep. Project No. 2342, Office of Hydropower Licensing, Division of Licensing and Compliance (2002). See p. 17 for "Economic Evaluation of Project Alternatives"
54. Fernandez, C., Leon, C., Steel, M., Vazquez-Polo, F.: Bayesian analysis of interval data contingent valuation models and pricing policies. J. Bus. Econ. Stat. **22**, 431–442 (2004)
55. Fink, A.: How to Ask Survey Questions. SAGE, Thousand Oaks and Wiley, New York (1985)
56. Fisher, A.C., Hanemann, M.W.: Quasi-option value: some misconceptions dispelled. J. Environ. Econ. Manage. **14**, 183–190 (1987)
57. Florio, M. (ed.): Cost-Benefit Analysis and Incentives in Evaluation. The Structural Funds of the European Union. Edward Elgar, Cheltenham (2007)
58. Førsund, F.R.: Hydropower Economics. Springer, New York (2007)
59. Førsund, F.: Energy in a bathtub: electricity trade between countries with different generation technologies. In: Johansson, P.O., Kriström, B. (eds.) Modern Cost-Benefit Analysis of Hydropower Conflicts, pp. 76–96. Edward Elgar, Cheltenham (2011)
60. Førsund, F., Hjalmarsson, L.: Renewable energy expansion and the value of balance regulation power. In: Johansson, P.O., Kriström, B. (eds.) Modern Cost-Benefit Analysis of Hydropower Conflicts, pp. 97–126. Edward Elgar, Cheltenham (2011)
61. Freeman III, A.M.: The measurement of environmental and resource values: theory and methods. Resources for the Future, Washington, DC (2003)
62. Fujiwara, D., Campbell, R.: Valuation techniques for social cost-benefit analysis: stated preference, revealed preference and subjective well-being approaches. A discussion of the current issues. HM Treasury, Department for Work and Pensions (2011)
63. Garthwaite, P., Kadane, J., O'Hagan, A.: Elicitation. Tech. Rep., Open University. http://statistics.open.ac.uk/TechnicalReports/Handbook-Elicitation.pdf (2004)
64. Goulder, L.H., Williams, R.C.: The substantial bias from ignoring general equilibrium effects in estimating excess burden, and a practical solution. J. Pol. Econ. **111**, 898–927 (2003)
65. Graham, D.A.: Cost-benefit analysis under uncertainty. Am. Econ. Rev. **71**, 715–725 (1981)
66. Green, R.J.: Electricity and markets. Oxf. Rev. Econ. Policy **21**, 67–87 (2005)
67. Green, G.P., O'Connor, J.P.: Water banking and restoration of endangered species habitat: an application to the Snake River. Contemp. Econ. Policy **19**(2), 225–237 (2001)
68. Greenberg, M.: Environmental Policy Analysis and Practice. Rutgers University Press, New Brunswick (2007)
69. Guesnerie, R.: General statements of second-best Pareto optimality. J. Math. Econ. **6**, 169–194 (1979)
70. Gumbel, E.J.: Statistics of Extremes. Columbia University Press, New York (1958)

71. Haab, T., McConnell, K.: Valuing environmental and natural resources. Edward Elgar, Cheltenham (2003)
72. Håkansson, C.: Cost-benefit analysis and valuation uncertainty. Ph.D.-Thesis, Department of Forest Economics, SLU (2007)
73. Hanley, N., Black, A.: Cost-benefit analysis of the water framework directive in Scotland. Integr. Environ. Assess. Manage. 2(2), 156–165 (2006)
74. Hanley, N., Craig, S.: Wilderness development decisions and the Krutilla-Fisher model: the case of scotland's flow country. Ecol. Econ. 4(2), 145–164 (1991)
75. Hanley, N., Spash, C.L.: Cost-benefit analysis and the environment. Edward Elgar, Cheltenham (1993)
76. Hanley, N., Kriström, B., Shogren, J.: Coherent arbitrariness: on value uncertainty for environmental goods. Land Econ. 85(29), 41–50 (2009)
77. Hansson, I.: Marginal cost of public funds for different tax instruments and government expenditures. Scand. J. Econ. 86, 115–130 (1984)
78. Hansson, H., Larsson, S.E., Nyström, O., Olsson, F., Ridell, B.: El från nya anläggningar. Tech. Rep. 07:50, Elforsk (2007)
79. Harrison, G., Vinod, H.: The sensitivity analysis of applied general equilibrium models: completely randomized factorial sampling designs. Rev. Econ. Stat. 74, 357–362 (1992)
80. Hartwick, J.M.: Natural resources, national accounting and economic depreciation. J. Public Econ. 43, 291–304 (1990)
81. Hatch, L.U., Hanson, T.R.: Change and conflict in land and water use: resource valuation in conflict resolution among competing users. J. Agric. Appl. Econ. 33(2), 297–306 (2001)
82. Haveman, R.H.: Water Resource Investment and the Public Interest. An Analysis of Federal Expenditures in Ten Southern States. Vanderbilt University Press, Nashville (1965)
83. Haveman, R.H.: The Economic Performance of Public Investments: An Ex Post Evaluation of Water Resources Investments. Johns Hopkins University Press, Baltimore and London (1972)
84. Heal, G., Kriström, B.: National income and the environment. In: Mäler, K.-G., Vincent, J.R. (eds.) Handbook of Environmental Economics. North-Holland, Amsterdam (2004)
85. Henry, C.: Investment decisions under uncertainty: the irreversibility effect. Am. Econ. Rev. 64, 1006–1012 (1974)
86. HM Treasury: The Green Book. Appraisal and Evaluation in Central Government. HMSO, London (2003). http://greenbook.treasury.gov.uk/index.htm
87. Hole, A.R.: A comparison of approaches to estimating confidence intervals for willingness to pay. Health Econ. 16, 827–840 (2007)
88. Hotelling, H.: Letter of June 18, 1947, to Newton B. Drury. Tech. Rep., Included in the report The Economics of Public Recreation: An Economic Study of the Monetary Evaluation of Recreation in the National Parks, 1949. Mimeographed. Land and Recreational Planning Division, National Park Service, Washington, DC (1947)
89. Huang, P.: Asymptotic properties of nonparametric estimation based on partly interval-censored data. Stat. Sin. 9, 501–519 (1999)
90. Hufschmidt, M., Fiering, M.: Simulation Techniques for Design of Water-Resource Systems. Macmillan, London (1966)
91. IEAB: Artificial production review economics analysis, phase i. http://www.nwcouncil.org/fw/ieab/Default.htm (2002)
92. IEAB: Juvenile passage cost effectiveness analysis for the Columbia River Basin: description and preliminary analysis. http://www.nwcouncil.org/fw/ieab/Default.htm (2004)
93. IEAB: Scoping for feasibility of Columbia River mainstream passage cost effectiveness analysis. http://www.nwcouncil.org/fw/ieab/Default.htm (2004)
94. IEAB: Review of: the estimated economic impacts of salmon fishing in Idaho. http://www.nwcouncil.org/fw/ieab/Default.htm (2005)
95. Jaeger, W.K., Mikesell, R.: Increasing streamflow to sustain salmon and other native fish in the Pacific Northwest. Contemp. Econ. Policy 20(4), 366–380 (2002)
96. Jammalamadaka, S., Mangalam, V.: Nonparametric estimation for middle-censored data. J. Nonparametr. Stat. 15, 253–265 (2003)

97. Jehle, G.A., Reny, P.J.: Advanced Microeconomic Theory. Addison-Wesley, Boston (2001)
98. Johansson, P.O.: The Economic Theory and Measurement of Environmental Benefits. Cambridge University Press, Cambridge, UK (1987)
99. Johansson, P.O.: Introduction to Modern Welfare Economics. Cambridge University Press, Cambridge, UK (1991)
100. Johansson, P.O.: Cost-Benefit Analysis of Environmental Change. Cambridge University Press, Cambridge, UK (1993)
101. Johansson, P.O.: Evaluating Health Risks. An Economic Approach. Cambridge University Press, Cambridge, UK (1995)
102. Johansson, P.O.: On the treatment of taxes in cost-benefit analysis. Cuad Econ. **80**, 149–161 (2010)
103. Johansson, P.O., Kriström, B.: A note on cost-benefit analysis, the marginal cost of public funds, and the marginal excess burden of taxes. Environ. Econ. **1**, 72–77 (2010)
104. Johansson, P.O., Kriström, B.: Comment on Burgess and Zerbe: on bank market power and the social discount rate. J. Benefit-Cost Anal. **2**(3), Article 6 (2011)
105. Johansson, P.O., Kriström, B.: The new economics of evaluating water projects. Annu. Rev. Resour. Econ. **3**, 1–21 (2011)
106. Johansson, P.O., Kriström, B.: On a new approach to social evaluations of environmental projects. Centre for Environmental and Resource Economics, Umeå University and the Swedish University of Agricultural Sciences, WP 2012–4, (2012)
107. Johansson, P.O., Kriström, B., Nyström, K.: On the evaluation of infrastructure investments: the case of electricity generation. Invited paper presented at the Eight Milan European Economy Workshop, 11–12 June, 2009. Published as Working Paper No. 2009–24, Department of Economics and Statistics, University of Milan (2009)
108. Jones-Lee, M.W.: Altruism and the value of other people's safety. J. Risk Uncertain. **4**, 213–219 (1991)
109. Kanninen, B.J. (ed.): Valuing environmental amenities using stated choice studies. A common sense approach to theory and practice. Springer, Dordrecht (2007)
110. Karp, L.: Global warming and hyperbolic discounting. CUDARE Working Paper 934R, Department of Agricultural & Resource Economics, UCB (2004)
111. Kneese, A.: Water resources research: a personal perspective. In: Smith, V. (ed.) Environmental Resources and Applied Environmental Economics: Essay in Honor of John V. Krutilla, pp. 45–55. Resources for the Future, Washington, DC (1988)
112. Kneese, A.: Whatever happened to benefit-cost analysis? Water Resour. Updat. **116**, 58–61 (2000)
113. Kokonendji, C.C., Zocchi, S.S.: Extensions of discrete triangular distributions and boundary bias in kernel estimation for discrete functions. Stat. Probab. Lett. **80**, 1655–1662 (2010)
114. Kolm, S.C., Ythier, J.M. (eds.): Handbook of the Economics of Giving, Altruism and Reciprocity Foundations, vol. 1. North-Holland, Amsterdam (2006)
115. Kolstad, C.D.: Environmental Economics, 2nd edn. Oxford University Press, Oxford (2010)
116. Kotchen, M.J., Moore, M.R., Lupi, F., Rutherford, E.S.: Environmental constraints on hydropower: an ex post benefit-cost analysis of dam relicensing in Michigan. Land Econ. **82**, 384–403 (2006)
117. Kriström, B.: W. Stanley Jevons (1888) on option value. J. Environ. Econ. Manage. **18**, 86–87 (1990)
118. Kriström, B., Laitila, T.: Stated preference methods for environmental valuation: a critical look. In: Folmer, H., Tietenberg, T. (eds.) The International Yearbook of Environmental and Resource Economics 2003/2004. A Survey of Current Issues. Edward Elgar, Cheltenham (2003)
119. Krutilla, J.V.: Conservation reconsidered. Am. Econ. Rev. **57**, 777–786 (1967)
120. Krutilla, J.V., Eckstein, O.: Multiple Purpose River Development: Studies in Applied Economic Analysis. Johns Hopkins University Press, Baltimore (1958)
121. Krutilla, J.V., Fisher, A.C.: The Economics of Natural Environments. Johns Hopkins University Press, Baltimore (1975)

122. Kynn, M.: The 'heuristics and biases' bias in expert elicitation. J. R. Stat. Soc. A **171**, 239–264 (2007)
123. Layard, R., Muyraz, G., Nickell, S.J.: The marginal utility of income. J. Public Econ. **92**, 1846–1857 (2008)
124. Li, C.Z., Löfgren, K.G.: Dynamic cost-benefit analysis of large projects: the role of capital cost. Econ. Lett. **109**, 128 – 130 (2010)
125. Ljusnans vattenregleringsföretag: (2008)
126. Loomis, J.B.: The economic value of instream flow: methodology and benefit. J. Environ. Manage. **24**, 169–179 (1987)
127. Loomis, J.B.: Quantifying recreation use values from removing dams and restoring free-flowing rivers: a contingent behavior travel cost demand model for the Lower Snake River. Water Resour. Res. **38**(6), 1066–1072 (2002)
128. Loomis, J.: Updated outdoor recreation use values on national forests and other public lands. Gen. Tech. Rep. PNW-GTR-658, U.S. Department of Agriculture, Forest Service, Pacific Northwest Research Station (2005)
129. Loomis, J.: A review of use and passive use values and an example: micro-meta nalysis of the contingent visitation benefits of removing dams. In: Johansson, P.O., Kriström, B. (eds.) Modern Cost-Benefit Analysis and Hydropower, pp. 100–120. Edward Elgar, Cheltenham (2011)
130. Louviere, J., Hensher, D., Swait, J., Adamowicz, W.: Stated Choice Methods. Cambridge University Press, Cambridge, UK (2000)
131. Maass, A., Hufschmidt, M.M., Dorfman, R., Thomas, H.A., Harold, A., Marglin, S.A., Fair, G.M.: Design of Water Resource Systems: New Techniques for Relating Econoimic Objectives, Engineering Analysis, and Governmental Planning. Harvard University Press, Cambridge, MA (1962)
132. Mäler, K.G.: National accounts and environmental resources. Environ. Resour. Econ. **1**, 1–15 (1991)
133. Manski, C., Molinari, F.: Rounding probabilistic expectations in surveys. J. Bus. Econ. Stat. **28**, 219–231 (2010)
134. Manski, C., Tamer, E.: Inference on regressions with interval data on a regressor or outcome. Econometrica **70**, 519–546 (2002)
135. Mas-Colell, A., Whinston, M.D., Green, J.R.: Microeconomic Theory. Oxford University Press, Oxford (1995)
136. McFadden, D., Bemmaor, A., Caro, F., Dominitz, J., Jun, B., Lewbel, A., Matzkin, R., Molinari, F., Schwarz, N., Willis, R., Winter, J.: Statistical analysis of choice experiments and surveys. Market Lett. **16**, 183–196 (2005)
137. Mensink, P., Requate, T.: The Dixit-Pindyck and the Arrow-Fisher-Hanemann-Henry option values are not equivalent: a note on Fisher (2000). Resour. Energy Econ. **27**, 83–88 (2005)
138. Mitchell, R., Carson, R.: Using Surveys to Value Public Goods: The Contingent Valuation Method. Resources for the Future, Washington, DC (1989)
139. Mooney, S.: A cost effectiveness analysis of actions to reduce stream temperature: a case study of the Mohawk Watershed. Ph.D. Dissertation, Department of Agricultural and Resource Economics, Oregon State University, Corvallis (1997)
140. Myles, G.D.: Public Economics. Cambridge University Press, Cambridge, UK (1995)
141. Narayanan, R.: Evaluation of recreation benefits of instream flows. J. Leis. Res. **18**(2), 116–128 (1986)
142. Navrud, S.: Pricing the European Environment. Oxford University Press, Oxford (1992)
143. Navrud, S.: Assessment of environmental valuation reference inventory (evri) and the expansion of its coverage to the eu, parts i, ii and iii. Tech. Rep., European Commission, DG Environment. http://europa.eu.int/comm/environment/enveco/studies2.htm#24 (1999)
144. NOAA: Final economic analysis of critical habitat designation for seven west coast salmon and steelhead ESUs. NOAA Fisheries (2005)

145. NPCC: Third annual report to the Northwest Governors on expenditures of the Bonneville Power Administration to implement the Columbia River Basin fish and wildlife program of the Northwest Power and Conservation Council, 1978– 2002. Council Document 2004-3, (2004)

146. Olsen, D., Richards, J., Scott, R.D.: Existence and sport values for doubling the size of Columbia River Basin salmon and steelhead runs. Rivers **2**(1), 44–56 (1991)

147. Olsson, M.: Optimal regulating power market bidding strategies in hydropower systems. Licentiate Thesis, Royal Institute of Technology, Department of Electrical Systems (2005)

148. Paulrud, A.: Economic valuation of sport-fishing in Sweden. Doctoral Disseration, Department of Forest Economics, SLU (2004). Acta Universitatis agriculturae Suecia. Silvestria vol. 323.

149. Pearce, D., Atkinson, G., Mourato, S.: Cost-benefit analysis and the environment. Recent developments. Organization for Economic Co-operation and Development, Paris (2006)

150. Pendleton, L., Mendelsohn, R.: Estimating recreation preferences using hedonic travel cost and random utility models. Environ. Resour. Econ. **17**, 89–108 (2000)

151. R Development Core Team: R: A language and environment for statistical computing. R Foundation for Statistical Computing, Vienna (2008). http://www.R-project.org. ISBN 3-900051-07-0

152. Ramsey, F.: A contribution to the theory of taxation. Econ. J. **37**, 47–61 (1927)

153. Reuss, M.: Water resorces, people and issues: interview with Professor Arthur Mass. Office of History, United States Army Corps of Engineers (1989)

154. Rivinoja, P., Gyllenhammar, A., Leonardsson, K.: Predicting European Grayling and Brown Trout densities at various flow scenarios in a regulated section of river Ljusnan, Sweden. Mimeo, Department of Wildlife, Fish, and Environmental studies, SE-901 83, Umeå, (2009)

155. Rivinoja., P., Gyllenhammar, A., Leonardsson, K.: Methods for predicting future salmonid population densities at various flow conditions applied in the river ljusnan. Tech. Rep. 3, Department of Wildlife, Fish, and Environmental studies, Swedish University of Agricultural Sciences, Umeå (2010)

156. Roggenkamp, M., Boisseleau, F.: The Regulation of Power Exchanges in Europe. Intersentia, Amsterdam (2005)

157. Russell, R., Baumann, D.: The Evolution of Water Resource Planning and Decision Making. Edward Elgar, Cheltenham (2009)

158. Sadler, H., Bennett, J., Reynolds, I., Smith, B.: Public choice in Tasmania: aspects of the lower Gordon River hydro-electric development proposal. Centre for Resource and Environmental Studies, Australian National University, Canberra (1980)

159. Saltveit, S.J.: The effects of stocking Atlantic salmon, Salmo salar, in a Norwegian regulated river. Fish. Manage. Ecol. **13**(3), 197–2005 (2006)

160. Schwartz, N.: Self reports: how the questions shape the answers. Am. Psychol. **54**, 93–105 (1999)

161. Sjöberg, P.: Essays on performance and growth in Swedish banking. Doctoral Disseration, Department of Economics, School of Business, Economics and Law, Göteborg University, No. 168. http://gupea.ub.gu.se/handle/2077/7619 (2007)

162. Smith, V.K.: Technical Change, Relative Prices and Environmental Resource Evaluation. Resources for the Future, Washington, DC (1974)

163. Smith, V.K.: Cost-benefit analysis and environmental policy: a comment. Kyklos **30**(2), 310–313 (1977)

164. Smith, V.K., Desvouges, W.: Measuring Water Quality Benefits. Kluwer, Amsterdam (1986)

165. Smith, V.K., Moore, E.M.: Does behavioral economics have a role for benefit cost analysis? In: Johansson, P.O., Kriström, B. (eds.) Modern Cost-Benefit Analysis of Hydropower Conflicts, pp. X–Y. Edward Elgar, Cheltenham (2011)

166. Statens energimyndighet: Långsiktsprognos 2008. ER 2009:14 (2009)

167. Stern, N.: The Economics of Climate Change. The Stern Review. Cambridge University Press, Cambridge, UK (2007)

168. Streiner, C., Loomis, J.B.: Estimating the benefits of urban stream restoration using the hedonic price method. Rivers **5**(4), 267–278 (1995)
169. Svenska Kraftnät: Storskalig utbyggnad av vindkraft. Konsekvenser för stamnätet och behovet av reglerkraft. Government report (2008)
170. Swedenergy: Handbook for the Swedish electricity market. Routines and exchange of information for trading and settlement. Swedenergy, Stockholm (2005). Translated version of issue 2.3, 2005-05-16, of Svensk elmarknadshandbok. Rutiner och informationsstruktur
171. Turnbull, B.W.: The empirical distribution function with arbitrarily grouped, censored and truncated data. J. R. Stat. Soc. **B 38**, 290–295 (1976)
172. Tversky, A., Kahnemann, D.: Judgment under uncertainty: heuristics and biases. Science **185**, 1124–1130 (1974)
173. U.S. Bureau of Reclamation: Operation of Glen Canyon Dam. Colorado River Storage Project, Arizona. Final environmental impact statement, The Bureau, Salt Lake City, March 1995
174. USFWS: Economic analysis of critical habitat designation for three populations of Bull Trout. Prepared by Northwest Economic Associates (2005)
175. Valentim, J., Prado, J.M.: Social discount rates. Tech. Rep., SSRN.com. http://ssrn.com/abstract=1113323 (2008)
176. Vattenfall: Bilaga till Vattenfall AB Elproduktion Nordens certifierade miljövarudeklaration EPD för el från Vattenfalls vattenkraft i Norden (2005)
177. Vattenfall: Annual Report. http://www.vattenfall.com/www/vf_com/vf%_com/369431inves/369463finan/481881annua/7118572006/index.jsp (2006)
178. Weigt, H.: Germany's wind energy: the potential for fossil capacity replacement and cost saving. Electricity Markets Working Papers WP-EM-29, Dresden University of Technology (2008)
179. Weinstein, M., Zeckhauser, R.: Critical ratios and efficient allocation. J. Public Econ. **2**, 147–157 (1973)
180. Weisbrod, B.: Collective-consumption services of individual-consumption goods. Q. J. Econ. **78**, 471–477 (1964)
181. Weitzman, M.L.: On the welfare significance of national product in a dynamic economy. Q. J. Econ. **90**, 156–162 (1976)
182. Weitzman, M.L.: A review of the Stern review on the economics of climate change. J. Econ. Lit. **45**, 703–724 (2007)
183. Welsh, M.P., Bishop, R.C., Baumgartner, R.M., Phillips, M.L.: Pilot test non-use value study. Glen Canyon Environmental studies. Tech. Rep. HBRD Inc., Madison (1994)
184. Welsh, M.P., Bishop, R.C., Phillips, M.L., Baumgartner, R.M.: Glen Canyon dam, Colorado River storage project, Arizona: nonuse values study final report. Tech. Rep., U.S. Bureau of Reclamation, Upper Colorado Regional Office, Salt Lake City (1997)
185. Willis, K., Nelson, G., Bye, A., Peacock, G.: An application of the Krutilla-Fisher model to appraising the benefits of green belt preservation versus site development. J. Environ. Plann. Manage. **36**, 73–90 (1993)
186. Wilson, R.: Non-linear Pricing. Oxford University Press, Oxford (1993)
187. Winter, J.: Response bias in survey-based measures of household consumption. Econ. Bull. **3**, 1–12 (2004)
188. Yu, J., Ranneby, B.: Estimation of wtp with point and self-selected interval responses. In: Johansson, P.O., Kriström, B. (eds.) Modern Cost-Benefit Analysis of Hydropower Conflicts, pp. 65–75. Edward Elgar, Cheltenham (2011)
189. Zerbe Jr., R.O.: The legal foundations of cost-benefit analysis. Charleston Law Rev. **2**, 93–184 (2007)
190. Zhao, X., Zhao, Q., Sun, J., Kim, J.: Generalized log-ranktests for partly interval-censored failure time data. Biom. J. **50**, 375–385 (2008)

Index